Flight Research at Ames

Fifty-Seven Years of Development and
Validation of Aeronautical Technology

NASA/SP-1998-3300

I0473681

Paul F. Borchers
James A. Franklin
and Jay W. Fletcher

National Aeronautics and
Space Administration

Ames Research Center
Moffett Field, California

Acknowledgments

The authors would like to recognize several individuals for their critical review, comments, and suggestions on the contents of this report. They are Wally Acree, Seth Anderson, Rod Bailey, Chris Blanken, Bill Bousman, Dick Bray, Jack Brilla, Luigi Cicolani, Greg Condon, Woody Cook, George Cooper, Lloyd Corliss, Brent Creer, Jeff Cross, Dallas Denery, Fred Drinkwater, John Dusterberry, Kip Edenborough, Dave Few, John D. Foster, John V. Foster, Ron Gerdes, Harry Goett, Bill Harper, Loran Haworth, Bill Hindson, Bob Jacobsen, Marty Maisel, Walt McNeill, Dennis Riddle, Stew Rolls, Mel Sadoff, Fred Schmitz, Rick Simmons, Barry Smith, Bill Snyder, Howard Turner, Del Watson, Maurie White, Brad Wick, and Zoltan Szoboszlay. The following individuals participated in the editorial review and production of this work and made significant contributions to the quality and style of the document: Lynn Albaugh, Roger Ashbaugh, Charlotte Barton, James Donald, Lloyd Popish, and Marie Westfall. The lead author, Paul Borchers, is credited for his considerable effort in initiating this document and for conducting an extensive search through various sources to compile the grist from which this report was eventually written. Following his departure from Ames, the co-authors assumed responsibility for completing this work.

AVAILABLE FROM:

NASA Center for AeroSpace Information
7121 Standard Drive
Hanover, MD 21076-1320
(301) 621-0390

National Technical Information Service
5285 Port Royal Road
Springfield, VA 22161
(703) 487-4650

Dedication

To the men and women who led the way in flight research at
Ames Research Center for six decades.

Contents

Introduction

One can get a proper insight into the practice of flying only by actual flying experiments.

Otto Lilienthal, 1896

Flight research has been an integral and essential part of the missions of, first, the National Advisory Committee for Aeronautics (NACA) and, later, its successor, the National Aeronautics and Space Administration (NASA).[1] The imperative of flight research was recognized from the outset in the NACA's charter: "[T]o supervise and direct the scientific study of the problems of flight with a view to their practical solution. . . ." The NACA's Ames Aeronautical Laboratory was established at Moffett Field, California, in 1939. Moffett Field was chosen as the site of the new laboratory for several reasons, including its predominantly good flying weather, moderate temperatures, and low air traffic density. The first building of the new laboratory, completed in August 1940, provided hangar and office space for flight research, as well as space for the management and administrative staff of the new laboratory (fig. 1). Appropriately, the first research conducted at Ames was a flight experiment. That first study was completed in 1940.

In its role as an aeronautical research laboratory, Ames, from its inception, made the most of the linkage between exploratory and developmental testing in its wind tunnels and in flight. In most respects, early research benefited from and was

Figure 1
Original flight research hangar circa 1941.

[1] In 1958, when the NACA was superseded by the National Aeronautics and Space Administration, the Ames Aeronautical Laboratory's name was changed to Ames Research Center.

stimulated by a strong association between the activities carried out in the 7- by 10-foot and the 40- by 80-foot low-speed and in the 16-foot high subsonic speed wind tunnels. Later on, the desire to broaden the base of flight research and to generalize the results of control and guidance investigations would lead to the development of Ames' flight simulators and to their equally strong interplay with flight. These interactions between the key aeronautical facilities and their research groups were a strength of the laboratory and made it exceptional among the world's aeronautical research establishments.

The research carried out in flight had numerous technical areas of emphasis over the years, and most of the individual experiments can be categorized accordingly. Individual aircraft may have served several purposes and thus may appear connected with more than one area of research. These areas are identified in the narrative to follow as icing research; transonic model testing; aerodynamics research; flying qualities, stability and control, and performance evaluation; variable stability aircraft; gunsight tracking and guidance and control displays; in-flight thrust reversing and steep approach research; boundary-layer control research; short takeoff and landing (STOL) and vertical and short takeoff and landing (V/STOL) aircraft research; and rotorcraft research.

From the earliest days of Ames Aeronautical Laboratory until the creation of NASA, the focus of flight research was on military aircraft and their operations. Icing research and the earliest efforts in aerodynamics and flying qualities research occurred during World War II and were intended to aid in the design and operation of aircraft for the Army Air Corps and the Navy. From the war's end until the late 1950s, motivation for research came from the need to achieve ever higher performance and to advance the technology in wing aerodynamics. However, impediments associated with controllability had to be overcome to realize these performance gains. Further, improvements in performance and controllability at low speed were required for the approach and landing inasmuch as low-speed characteristics typically were adversely affected by high-speed enhancements. Variable stability aircraft were created as in-flight research facilities in order to expand upon the characteristics of a single configuration to acquire data for the development of flying qualities design criteria for individual categories of aircraft. The work in aerodynamics; flying qualities, stability and control, and performance; gunsight tracking and guidance and control; and in-flight thrust reversing was directed accordingly by the NACA, with advice and encouragement from the military services and the aircraft industry. Throughout this era, strong guidance was provided to the program through the NACA subcommittees, whose membership was drawn from the industry and military services.

Upon the transition from the NACA to NASA, these areas of research came to an abrupt halt. In mid-1959, Ames was directed by NASA headquarters to transfer all flight research to the Flight Research Center located at Edwards Air Force Base in southern California. Within 2 years, all Ames high-performance aircraft were moved to their new home in the high desert. However, headquarters assigned Ames the responsibility for powered-lift research, including flight research with STOL and V/STOL aircraft. This decision was influenced by Ames' broad technical background with this category of aircraft in aerodynamics, performance, stability and control, flying qualities, and operations and because of the presence of the 40- by 80-foot wind tunnel and its experienced aerodynamics staff that had developed considerable expertise in powered-lift technology. Another influence on this decision was the interest the U.S. Army had expressed in this area of technology and the beginnings of what would become a cooperative program in aeronautical research with Ames. Thus, powered-lift research grew into a major effort that has lasted to the present day, supporting military along with newly emerging civil needs. It included the development and flight of several proof-of-concept aircraft, particularly the XV-15 tilt rotor, which stands as one of Ames' most important contributions to aeronautical technology. Further, it was soon to be augmented with rotorcraft flight research when NASA chose to consolidate rotary-wing technology efforts at Ames in the late 1970s. This research was supported and strongly influenced by the Army through its research laboratory, which had been established and collocated at Ames in the late 1960s. This collaborative program continues to this day.

Throughout most of the period of flight research at Ames Aeronautical Laboratory until its transition to Ames Research Center under NASA, a few select individuals provided top leadership. John Parsons served initially as chief of the Full-Scale and Flight Research Division, primarily to oversee its establishment as a working research unit. Lawrence (Larry) Clousing headed up the Flight Research Section under Parsons and provided technical guidance for flight research during the war effort. Lewis A. (Lew) Rodert was in charge of the Flight Engineering Section. At the end of World War II, Parsons was succeeded as division chief by Harry Goett. Goett, along with Steven (Steve) Belsley, the chief of the Flight Research Branch, and William H. (Bill) McAvoy and George Cooper, successively chiefs of the Flight Operations Branch, provided crucial leadership for over a decade. They established a working atmosphere of individual excellence and dedication that was leavened with considerable irreverence for authority. Goett set the tone with his weekly meetings with the staff, which included intense attention to technical objectives and results, with a minimum of administrative encumbrances. He was also known for his low tolerance of high-level administrative interference in the work. An illustrative example was his abrupt termination of a telephone conversation with an official in the Washington office of the NACA (he hung up) while debating the advisability of a particular flight experiment. As division chief, he was in a position to encourage the interaction between the flight and wind tunnel testing noted earlier. Belsley, like Goett, demanded technical competency and integrity, and was never known for his diplomatic approach in discussing issues of consequence. Otherwise, his sonorous voice eliminated the need for an intercom in the hangar. McAvoy was well regarded for his contributions to hazardous test flights, and for his oversight of a highly capable staff of pilots, all of whom went on to distinguish themselves throughout their careers. Cooper, a soft-spoken person and gentleman through and through, led by example and showed considerable insight into a wide range of aeronautical technologies. To this day, he continues to be revered by his colleagues throughout the profession for his accomplishments and personal character, not to mention for his second career as an enologist and vintner. The NACA quality standards set by these men persisted in the organization through its subsequent evolutions of structure and leadership and influenced the content and conduct of the research. It has also been a source of pride to the individuals involved throughout the ensuing years.

Not long after the creation of NASA in 1958, Charles W. (Bill) Harper, another spirited sort who had learned the business under Harry Goett, succeeded Goett as division chief. Harper's first duty when he assumed this position was to deal with the headquarters directive to move Ames' research aircraft to the Flight Research Center at Edwards Air Force Base. His quick response in developing research in powered-lift technology provided the basis for establishing Ames' flight research role in this area. At about that same time, Seth Anderson took over as the research branch chief from Steve Belsley and held that position for another 10 years. During that time, he became widely known for his expertise and leadership in V/STOL dynamics and control. Starting in the early 1970s, the organization evolved into a much broader entity, with participation from two branches in the original division and from the Aircraft Projects Office set up within the Aeronautics Directorate. Additionally, flight operations became a much larger organization with the creation of a division structure within aeronautics. Key leadership at the research division level came from Bradford (Brad) Wick, then successively from C. Thomas (Tom) Snyder and Gregory Condon. The projects office was established and led initially by Woodrow (Woody) Cook. After Cook's retirement in the late 1970s, the direction of this organization, now a full division, came from Wallace (Wally) Deckert and David (Dave) Few. Research branch leadership in flight dynamics and control was provided by a succession of individuals who followed Seth Anderson in that position, first Maurice (Maurie) White, then James (Jack) Franklin, Victor Lebacqz, Edwin Aiken, and William (Bill) Hindson. Navigation and guidance was led first by Brent Creer and later by Dallas Denery. With the creation of the Flight Operations Division in the mid-1970s, which occurred after the retirement of George Cooper, division chiefs in succession were David Reese, Fred Drinkwater, and James (Jim) Martin. Flight operations branch leadership following Cooper came from Robert (Bob) Innis, Jim Martin, Warren Hall, and George Tucker. Other organizations supported this effort throughout, particularly those groups associated with aircraft maintenance, development of advanced systems, and data acquisition. In particular, the Aircraft Services and Inspection Branches, the Metal Fabrication Branch, the Flight Systems Branch, and the Instrument and Avionics Research Branches were essential to the success of the programs over the years. Ray Braig, superintendent of aircraft, who headed up aircraft maintenance at the outset, was particularly esteemed by his colleagues for his competence and hard work.

A variety of sources, both formal and informal, written and oral, have been used in preparing this history. For the first quarter century of flight research, Edwin Hartman's history of Ames (ref. 1) provides extensive material concerning the areas of icing research; transonic model testing; aerodynamics research; flying qualities, stability and control, and performance evaluation; variable stability aircraft; gunsight tracking and guidance and control; in-flight thrust reversing; and boundary-layer control. During the subsequent 30-plus years, further information on research in boundary-layer control, and on STOL and V/STOL aircraft and rotorcraft comes from the authors' personal experiences and recollections. We have drawn considerable information as well from several of our colleagues and from Ames alumni. The Ames technical library has also been a valuable source of formal NACA and NASA reports. The aircraft identifications that are applicable to research at Ames are generally taken from a list of NACA and NASA aircraft compiled by Robert L. Burns at NASA Goddard Space Flight Center. A set of notebooks kept by Donovan (Don) Heinle, a test pilot at Ames in the 1950s and early 1960s, listed all the aircraft that appeared at Ames and the projects in which they were involved. His notes also included a list of Ames research pilots and the dates of their first flights. In some cases, this is the only documentation available on the activities of the early aircraft. Unfortunately, Heinle's records ended when he lost his life while flying an F-101 Voodoo at Edwards Air Force Base. Additional information came from an inventory of photographs collected by the Flight Research Branch. These photographs cover the period from the first flights at Ames to the mid-1950s. All of these photographs had been stapled to cards, on which the subject of the photograph, the name of the project, and the date of the photograph were listed. Although they are not a complete history, many of the photographs offer insights into the nature of the flight experiments at the Center, the modifications that were made to the aircraft, the actual appearance of some of the rarer types of aircraft, the individuals involved, and the dates during which some of the flying took place.

The rationale for the various research areas is described in the narrative that follows, along with a brief accounting of some of the more prominent flight programs. No attempt is made to cover every aircraft tested at Ames. Rather, highlights of the programs are indicated, along with anecdotal descriptions of the individuals involved and some interesting results of the programs. Tables that provide identifying information about the aircraft flown in the various research programs and relevant photographs appear in each section. Electronic versions of the photographs are available and can be downloaded from the Ames Imaging Library located at *http://ails.arc.nasa.gov.*

Icing
Research

The first research program undertaken at the new Ames Aeronautical Laboratory concerned the development and testing of icing protection systems for military aircraft. The need of the military services to operate their new high-performance bombers and transports year around in adverse weather led the NACA to initiate this program. This effort had begun sometime earlier at the Langley Aeronautical Laboratory with wind tunnel and flight tests. The flight activity, including personnel, was relocated to Ames in 1940. The first flight experiment at Ames was carried out on the O-47A-1, originally an Army observation aircraft (fig. 2), to obtain initial results for this program. As noted in table 1, this was the first aircraft to arrive at the laboratory and, starting in September 1940, it served for a brief time as the research aircraft for the evaluation of heated-wing deicing. The following January, icing research continued with the arrival of the Lockheed 12A Electra (fig. 3), which had been involved in this research at Langley. The deicing concept involved circulation of hot exhaust gas from the engines through the wings, and it proved to be effective. The first research publication of the Ames Aeronautical Laboratory (ref. 2) covered this work. This system was further developed to use inducted free-stream air warmed by an exhaust heat exchanger; it was then applied to the B-17 and B-24 heavy bombers (figs. 4 and 5) for wing and empennage deicing. Flight tests were carried out during the winter of 1943 at the Army Air Force Icing Research Base in Minneapolis. Results published in reference 3 for the B-17 showed that the system worked; similar performance was achieved for the B-24. With that success, the system was used on production versions of the Navy's PBY Catalina.

Following the B-17 and B-24 programs, a Curtiss C-46A Commando (fig. 6) was developed as a dedicated flight research facility and was used extensively for testing advanced deicing techniques. The aircraft employed fuselage-mounted airfoil models at one time as a part of this work. Additionally, the aircraft was equipped with instrumentation to collect data to aid in predicting icing conditions. Operations were carried out in natural icing conditions without incident. Analytical predictions of deicing system requirements were found to be conservative and, when they were met, ice buildup on the wings was prevented (ref. 4). Propeller icing experiments were also conducted and measurements were obtained that showed the propeller efficiency loss caused by ice formation. Analyses of the effects on propeller efficiency showed qualitative agreement with the flight data (ref. 5). Later on, the BTD-1 Destroyer was also the subject of a deicing study, and electrically heated "gloves" were tested on the wings and tail of the P-38J Lightning (fig. 7). Ames' original staff of pilots involved in these flight tests appears in figure 8.

Figure 2
North American O-47A-1 with Bill McAvoy.

Figure 3
Lockheed 12A Electra.

Figure 4
Boeing B-17F Flying Fortress.

TABLE 1. AIRCRAFT USED FOR ICING RESEARCH		
Aircraft Name	Arrival or First Flight Date	Departure Date
O-47A-1 (AAC37-323)	September 5, 1940	March 13, 1946
Lockheed 12A (NC 17396 NACA 97)	January 20, 1941	1947
XB-24F-CO (AAF41-11678)	May 13, 1942	1942
B-17F (AAF42-5474)	August 28, 1942	November 19, 1943
C-46A (AAF41-12293)	March 10, 1943	March 24, 1949
BTD-1 (Bu. No. 04968)	July 28, 1944	June 30, 1947
P-38J (AAF43-28519)	August 30, 1944	March 15, 1946

Figure 5
Consolidated XB-24F-CO Liberator.

Figure 6
Curtiss C-46A Commando.

Figure 7
Lockheed P-38J Lightning.

Reflecting the major contributions made at Ames over the duration of this program, Lew Rodert received the Collier Trophy in 1947 for his leadership of the effort, and, in 1943, Bill McAvoy received the Octave Chanute Award for flight testing in dangerous icing conditions. Other significant contributors were Larry Clousing, who, along with McAvoy flew many hazardous test flights, Carr Neel, who made significant contributions to the design of the deicing system for the C-46 and developed an instrument to measure liquid water content in flight, and Alun Jones, who also contributed to the system design. Jones wrote the summary report of the C-46 work (ref. 6) that described the analysis of heat requirements for ice protection; the design, fabrication, and installation of the system; performance tests in operational icing conditions; and the evaluation of effects on cruise performance and structural integrity of the wings. It concluded that prediction methods were now adequate, though still conservative for system design, and that thermal effects on the wing structure could be avoided with proper design approach. Interest in icing research at Ames virtually ceased by 1949 as the Flight Engineering Branch shifted emphasis to transonic studies.

Figure 8
Ames research pilots circa 1942. From left to right: Larry Clousing, Bill McAvoy, Jim Nissen.

Transonic
Model Testing

Interest in the transonic flight regime increased markedly after the Second World War, reflecting further attempts to increase aircraft performance. However, wind tunnels of the time were inadequate for carrying out this kind of research. For a time, tests were conducted in flight with small airfoil and aircraft models.[2] These models were attached to the wings of conventional fighter aircraft, in some cases on a raised surface known as a "transonic bump," where the airflow would be compressed and accelerated to transonic speeds during dives (hence the term "wing-flow" testing). Transonic model test aircraft are noted in table 2. One P-51B (figs. 9 and 10), one P-51D (fig. 11), and two P-51H Mustangs were the primary aircraft used at Ames in wing-flow test flights. Larry Clousing performed the first of these tests in the P-51B. In experiments with a thin, straight wing with a symmetrical double-wedge profile, lift data matched theoretical and wind tunnel predictions up to Mach 0.82. For Mach numbers through the transonic range up to 1.2, trends in lift disagreed with theory. Pitching moment data indicated aerodynamic center movement approached the theoretical predictions for subsonic and supersonic flow (ref. 7). In another series of tests, the control effectiveness for several flat-plate delta-wing planforms with trailing-edge flaps was explored. Pitching moments were measured for various flap angles over a speed range up to Mach 1.1. Data showed a reasonable trend through the transonic range (ref. 8).

In another approach to acquiring transonic aerodynamic data, heavily weighted models of the configuration of interest were dropped from high altitudes. In those tests, which were conducted at Edwards Air Force Base, aerodynamic bodies that were to be evaluated in the transonic flight regime were released from an aircraft at altitudes up to 43,000 feet. The instrumented bodies would pass through the transonic speed range in free fall, during which they were oscillated through a range of angles of attack and were then decelerated and recovered by means of air brakes and parachutes. Testing at these altitudes was arduous and, although the pilots wore heavy flight suits, the model drops were made on the first run to reduce the pilots' exposure to the extreme cold.[3] The F-15A-1-NO aircraft (fig. 12), a reconnaissance model of the P-61 night fighter, was used for these tests. The high-altitude capability of the F-15A made it the ideal "mother ship" for this work. An aircraft similar to this one, an ERF-61C (fig. 13), owned by the Smithsonian Institution, was lent to Ames to be

Figure 9
North American P-51B Mustang.

Figure 10
Wing flow models on P-51B.

Figure 11
P-51D.

TABLE 2. AIRCRAFT USED FOR TRANSONIC MODEL TESTS		
Aircraft	**Arrival or First Flight Date**	**Departure Date**
P-51B (AAF43-12094)	November 16, 1944	September 9, 1947
P-51H (AAF44-64691)	January 25, 1947	May 17, 1948
P-51D (AAF44-74944)	April 15, 1947	November 23, 1949
P-51H (AAF44-64703 NACA 110)	November 6, 1947	May 17, 1956
F-15A-1-NO (AAF45-59300 NACA 111)	February 6, 1948	October, 1954
ERF-61C-1-NO (AAF43-8330 NACA 330, NACA 111)	February 5, 1951	August 10, 1954

[2] Harry Goett and Bill Harper 1998: personal communication.
[3] George Cooper 1998: personal communication.

Figure 12
Northrop F-15A-1-NO.

Figure 13
Northrop ERF-61C-1-NO.

used in this program as well. Pilots who participated in this work were George Cooper, Rudolph (Rudy) Van Dyke, Don Heinle, and Fred Drinkwater. As with the wing-flow tests, qualitative results were obtained; nevertheless, the advent of the new transonic tunnels supplanted flight testing as a means of documenting the aerodynamics of this flight regime. The air-brake and parachute systems developed for these tests were subsequently used by many agencies for rocket and satellite payload recovery. The NACA test pilots who were at Ames in 1949 are shown in figure 14.

Figure 14
Ames research pilots circa 1949. From left to right: Ray McPherson, Rudy Van Dyke, Bill McAvoy, George Cooper, Larry Clousing.

Aerodynamics Research

Early flight research that focused on aerodynamic issues was concerned with understanding drag, air loads, and compressibility phenomena that influenced both the performance and control of the aircraft at high speed. This work was motivated by problems uncovered in the design of these high-performance aircraft and in early operational experience with them. Furthermore, as Ames engineers and pilots gathered information on their own, additional ideas surfaced that suggested new approaches to solving these problems.[4] Many of the World War II and postwar aircraft involved are noted in table 3 and shown in figures 15–28. Members of the flight research section, as they appeared in 1946, are shown in figure 29. The new flight research hangar, under construction and soon to be occupied, can be seen in figure 30.

It was of particular interest to aerodynamicists that data be acquired from an aircraft in flight to use in validating wind tunnel measurements of drag. The first aircraft used in this effort was the P-51B Mustang (one of the first production aircraft to have laminar flow airfoils). To carry out this experiment without interference from the propeller slipstream, the propeller of the aircraft was removed and its oil and coolant ducts blocked so that it resembled the wind tunnel model. The aircraft, flown by James (Jim) Nissen, was towed aloft by a P-61 and released. Careful measurements of longitudinal deceleration were used to determine aircraft drag, and the pilot of the Mustang made a powerless, gliding landing. During one of these flights, the tow cables accidentally separated from the P-61 and wrapped themselves around the P-51, interfering with the pilot's control of the aircraft. Despite a crash landing, Nissen

Figure 15
Lockheed P-38F Lightning.

TABLE 3. AIRCRAFT USED FOR AERODYNAMICS RESEARCH		
Aircraft Name	**Arrival or First Flight Date**	**Departure Date**
P-38F (AAF41-7632)	December 30, 1942	July 16, 1943
P-51B (AAF43-12111)	August 11, 1943	September 7, 1947
P-39N	September 13, 1943	August 29, 1944
P-63A (AAF42-68892)	February 17, 1944	June 18, 1946
XSB2D-1 (Bu. No. 03552)	June 12, 1944	January 10, 1946
YP-80A (AAF44-83023)	September 19, 1944	January 27, 1947
P-51B (AAF43-12094)	November 16, 1944	September 9, 1947
P-63A-6 (AAF42-68941)	January 27, 1945	June 18, 1946
P-47D-25 (AAF42-26408)	April 27, 1945	September 7, 1947
P-80A-1 (AAF44-85299 NACA 131)	December 18, 1946	June 6, 1955
YP-84A-5 (AAF45-59488)	December 2, 1947	October 5, 1948
YF-84A (AAF)	February 14, 1949	December 20, 1950
F-86A (AAF48-291 NACA 116)	August 29, 1949	January 11, 1960
YF-93A (AF48-317 NACA 139, 48-318 NACA 151)	February 5, 1951 June 5, 1951	1953
YF-86D (AF 50-577 NACA 149)	June 26, 1952	February 15, 1960
F4D-1 (Bu. No. 134759)	April 4, 1956	October 16, 1959
F5D-1 (Bu. No. 139208a NASA 212)	August 20, 1957	January 16, 1961
Lear 23 (cn23-049 NASA 701)	September 17, 1965	January 11, 1980

Figure 16
P-51B.

Figure 17
P-51B towed by P-61A Black Widow.

[4] Harry Goett 1998: personal communication.

Figure 18
Bell P-39N Airacobra.

Figure 19
Douglas XSB2D-1 crashed in prune orchard in Sunnyvale, Calif.

Figure 20
Lockheed YP-80A Shooting Star.

survived without major injury. Comparison of flight results and data from the 16-foot wind tunnel are presented in reference 9; they show good agreement below the drag-rise Mach number. In flight, the drag rise occurred at lower Mach numbers than it did in the wind tunnel; it was speculated that aeroelastic deformation under flight loads was the cause. The NACA technical report is prefaced with an editorial note by the chairman of the NACA commending Nissen's "great skill and courage" in staying with the airplane to keep this crucial data from being lost. Nissen eventually left Ames, initially for North American, and then on to develop what was to become the San Jose airport. He ultimately managed the airport for the City of San Jose. In his later years he owned and flew a Curtiss Jenny and a Thomas-Morse Scout from the airfield at his home near Livermore, California.[5]

During and after World War II, a number of military aircraft were used in general investigations of high-speed flight phenomena. Because of structural failures in the tails of several fighter airplanes during high-speed dives, Ames undertook the measurement of tail loads on the Bell P-39N under a range of flight and maneuver conditions. Larry Clousing conducted most of those tests and demonstrated considerable courage in pushing the aircraft to its limits to obtain measurements of critical loads. Results of that work, examples of which are presented in references 10–12, pointed to deficiencies in the prediction methods, which not only underestimated loads but in some cases showed them to be in the opposite direction. Buffet loads were a contributing factor to tail loads and may have led to partial failure of the tail on one flight. Photographs from the tests showed that the fabric covering on the elevator bulged out at higher airspeeds, a condition that was followed by a partial structural failure of the horizontal stabilizer during the pullout from a dive. Analyses of the data established the effects of critical Mach number on the wing center of pressure and showed that the lift curve behaved as predicted. Other tests had shown the contribution of shock-induced wing stall to tail angle of attack and that influence, in turn, on a strong nose-down pitching moment. Distortion of the elevator fabric also served to move the stick free neutral point aft, thereby increasing the stick force gradient (ref. 13). Analytical predictions of vertical tail loads during rolling pullout maneuvers compared favorably with flight measurements for the airplane based on dynamic analysis of sideslip excursions during the maneuver (ref. 14).

Drag measurements were also performed on the P-39N-1, in which minimum drag and drag-rise Mach numbers were documented in tests carried out to Mach 0.8 (ref. 15). Engine thrust was estimated using propeller efficiency and engine horsepower predictions in order to extract drag from the data.

High-speed buffet was evaluated in dive tests with the P-51 by George Cooper. During these tests, Cooper observed sunlight refracting through the shock wave, identifying its presence on the wing and noting a correlation between its movement and the occurrence of buffeting. Cooper co-authored a report with George Rathert on the visual observation of these shock waves (ref. 16). At one time, one of the

[5] George Cooper 1998: personal communication.

P-51H aircraft was fitted with a Schlieren system for visualization of shock waves on the wing (ref. 17).

Strong nose-down pitching moments occurring at high speed limited pitch control for these high-performance aircraft. Wind tunnel tests indicated that slight upward deflection of the flaps could reduce those moments and expand the controllable flight envelope. At the suggestion of John Spreiter and Jim Nissen, flaps on the P-51H and F8F-1 were modified and a program was conducted on the two aircraft to substantiate the tunnel results. Maurie White and Melvin (Mel) Sadoff ran the tests, and George Cooper and Larry Clousing carried out the flights after Nissen had left Ames. The degree of elevator required to trim was reduced substantially for the P-51H up to Mach number 0.795, the highest speed tested. However, results from the F8F-1 were not encouraging; the favorable contribution realized on the P-51H through reduction of changes in tail angle of attack was offset by an increased nose-down pitching moment contribution from the wing (ref. 18).

Figure 21
Bell P-63A-6 Kingcobra.

Different propellers were tested on the XSB2D-1 during maximum power ground runs and in-flight performance evaluations. The aircraft was also flown so flight data and data obtained in tests run in the 40- by 80-foot wind tunnel could be compared. Welko Gasich reported the results of these tests and used them to compare alternative methods for calculating takeoff ground run (ref. 19). During what turned out to be the final flight of the XSB2D-1, an engine fire occurred, and George Cooper made a successful emergency landing, short of the main runway at Moffett Field and between tree rows in a Sunnyvale prune orchard. In the process, 84 trees were mowed down and the airplane's wings were severely cropped, but Cooper and Gasich, who flew on board as the test engineer, escaped without injury. The local farmer, who was personally acquainted with Cooper, was astonished to see George climb from the cockpit. Cooper, unflappable as usual, exclaimed to his friend, "You keep asking me to drop in on you sometime, so here I am."[6]

Figure 22
P-80A-1 showing vortex generator installation.

Cooper was also involved in flight tests of a reversible-pitch propeller on the P-47D Thunderbolt. These tests were performed to evaluate the handling characteristics of the aircraft with the propeller used as a brake during a dive. At the conclusion of one test, the propeller failed to return to its normal pitch setting. Cooper was almost forced to land with the aircraft in this condition, but fortunately the propeller snapped out of its reversed pitch setting while the aircraft was on approach. He added power, climbed away, and made a safe landing on the next approach.[7]

With the coming of the jet aircraft, compressible flow phenomena raised issues with aircraft performance and handling, leading to new demands for flight testing. The YP-80A was the first jet aircraft at Ames. The aileron buzz problem on the P-80 was of particular concern to its designers and became the focus of an intensive investigation in the 16-foot transonic tunnel as well as in the P-80A-1.[8] The Ames team consisted of Harvey Brown, George Rathert of the engineering staff, and Larry

Figure 23
Republic YP-84A-5.

---

[6] George Cooper 1998: personal communication.
[7] George Cooper 1998: personal communication.
[8] Seth Anderson 1998: personal communication.

Figure 24
North American F-86A Sabre.

Figure 25
North American YF-93A with NACA
submerged inlet.

Figure 26
North American YF-93A with scoop inlet.

Clousing, the principal test pilot. Clousing performed dive tests to the highest speeds yet achieved with the aircraft in pursuit of data to identify the source of the problem. In so doing, the aircraft expanded the transonic flight envelope to 0.866 Mach. Results reported in references 20 and 21 showed the effect of critical Mach number on the aileron oscillation and tied it to shock-induced separation on the upper surface that influenced aileron hinge moments. The data from the 16-foot wind tunnel gave a good indication of the onset conditions. The phenomena had been observed earlier in P-39N dives and the correlation with Mach number was noted at that time; the oscillating shock was identified as the cause (ref. 22). Flight tests of these aircraft also investigated boundary-layer characteristics and removal, which is important for maintaining the proper airflow to fuselage-mounted jet engines. The YP-80A was also used for tail pipe temperature measurements. Clousing's contribution to these tests, as well as to the earlier testing of the P-39, was due to his skill and courage as a test pilot and to his interpretation of the results and the test techniques involved. For the benefit of his fellow pilots and engineers, he published a report (ref. 23) that reflected on the hazards of high-speed testing with these aircraft.

John Spreiter led a series of flight tests using several aircraft to determine the effect of Mach number and Reynolds number on maximum lift for comparison with results from the 16-foot tunnel. These tests were made on the YP-80A and on five propeller-driven fighters, the P-38F, P-39N, F6F-3, P-51B, and the P-63A. The data correlation was encouraging except when buffeting limited the angle of attack that could be achieved in flight (ref. 24).

George Cooper and Rudy Van Dyke began flight tests of the Air Force's new F-86A Sabre in 1949. They made prolonged dives, starting from 46,000 feet, in which the F-86A reached very high speeds. These flights opened up the aircraft's supersonic envelope and preceded North American and Air Force tests of the aircraft at these speeds. At about the same time, people in the general area began to hear explosions that occurred without any apparent reason. Eventually, these "explosions" were correlated with the dive tests of the F-86 Sabre; they occurred when the aircraft reached supersonic speeds. This was the first time the "sonic boom" phenomenon had been associated with the supersonic flight of an aircraft. [9] It is also noteworthy that these two pilots were routinely breaking the sound barrier at a time when only a small number of others, based primarily at Muroc Dry Lake, had done the same thing.

The F-86A was known to have problems with pitch-up and lateral control during transonic flight, and a flight program was carried out to document the aircraft's longitudinal stability characteristics for comparison with wind tunnel tests (ref. 25). Two F-86s were then flight tested to assess the effects of the height of the horizontal stabilizer on pitch stability and to check earlier wind tunnel data (ref. 26). It was found that the two tail locations tested had no bearing on the pitch-up characteristics. A more significant factor was the use of servo-powered longitudinal controls on one airplane, which eliminated the stick-free instability that characterized the original reversible controls. Comparisons of wind tunnel and flight data were inconsistent;

[9] George Cooper 1998: personal communication.

these inconsistencies were attributed to detailed differences between the tunnel models and the aircraft and, as well, to the differences in Reynolds number. In further tests, Seth Anderson and Frederick Matteson investigated alterations of the wing leading edge to determine the influence on pitch-up. Leading-edge camber was found to have no effect on longitudinal stability, just as wind tunnel testing had indicated. Modest increases in maximum lift were observed, although abrupt asymmetric stall was experienced until flow fences were installed (ref. 27). Partial-span leading-edge chord extensions did eliminate pitch-up below Mach 0.84, but had no comparable influence at higher Mach numbers (ref. 28). Mel Sadoff, John Stewart, and George Cooper performed studies to correlate pilot opinion with the pitch-up characteristics of several aircraft. Using data from the F-84F, F-86A, D, and F models, and F-100 fighters and the B-47 bomber, they found that pilot opinion could be related to angle of attack and normal acceleration overshoots. Pitch-up tendencies were noted to range from mild to severe and, in the worst cases, design load factors and tail loads were exceeded for some of these airplanes (ref. 29).

Figure 27
North American YF-86D showing vortex generator installation.

Figure 28
Douglas F4D-1 Skyray with Don Heinle, Stew Rolls, and Walter Liewar.

Figure 29
Flight Research Section circa 1946. Front row: Bob Bishop, Mary Anderson, Helen Brummer, Mary Thompson, Larry Clousing, Chan Cathcart. Second row: Ben Gadeberg, Carl Stough, Bill Turner, George Galster, Betty Adams, Bob Reynolds. Third row: Stew Rolls, George Rathert, John Spreiter, Tom Keller, Mel Sadoff, Kinsenger, Paul Steffen. Fourth row: Welko Gasich, Maurie White, Steve Belsley, Bill Kauffman, Seth Anderson, Carl Hanson, Harvey Brown. Not pictured: Howard Turner, Dick Skoog, Gavras, Don Christopherson, Bunnel.

Figure 30

New flight research hangar under construction circa 1945.

Figure 31

Ames Research Center wind tunnels.

Lateral stability and control tests on the F-86 showed that wing dropping was caused by a directional asymmetry, and an abrupt increase in dihedral effect accompanied by a decrease in roll-control effectiveness (ref. 30). Later, the F-86A and D were both flown with vortex generators in attempts to improve longitudinal and lateral control; the vortex generators proved successful in alleviating these problems (ref. 31). Flow fences were also installed and tested on the F-86A and they too were shown to reduce the pitch-up tendency at transonic speeds (ref. 32). Pitch-up tendencies of the YP-84A-5 were also evaluated.

The YF-93A aircraft was the first to use flush NACA engine inlets. The flush inlet design had undergone extensive development in the 7- by 10-foot, 40- by 80-foot, and 16-foot wind tunnels (fig. 31) under the guidance of Emmett Mossman. North American developed two YF-93A prototype aircraft from the F-86 Sabre design under Air Force sponsorship. One aircraft (AF 48-317) was built with flush inlets; the other had conventional scoop inlets. Two interchangeable tail sections were provided as well.[10] The NACA acquired both aircraft, and Stewart (Stew) Rolls conducted flight tests of them to compare the two inlet designs and to check results against data from the wind tunnel tests used in their development. Measurements were made of inlet pressure recovery and overall airplane drag of the aircraft. Flight data showed that the submerged inlet had higher pressure recovery and higher drag than the scoop inlet below Mach 0.89, with overall performance essentially the same between the two. Sealing the boundary-layer bleeds improved the performance of both inlets (ref. 33).

Thrust measurements were made on the YF-93A by Stew Rolls using a movable pitot-static and temperature probe in the jet exhaust. Gross thrust and airflow at the jet exit were obtained with the thrust measurement accuracy determined to be within 5 percent at full power (ref. 34).

The general performance characteristics of the F4D-1 aircraft were examined at Ames. One report on these tests (ref. 35) presents a thorough analysis of minimum drag and drag due to lift for this tailless delta wing configuration and compares the results with data from the 14-foot transonic wind tunnel. Tunnel measurements for minimum drag were generally lower than those obtained in flight, a discrepancy attributed to lack of detail in the wind tunnel model.

Ames had been conducting research on leading-edge vortices for low aspect ratio swept wings and, in response to discussions with French officials concerning a new wing planform for a supersonic transport, a program was initiated in the 40- by 80-foot wind tunnel to investigate the application of the so called Ogee planform to the F5D-1 aircraft.[11] The sharp and highly swept inboard portion of the leading edge produced a strong vortex that was shown in the wind tunnel tests to stabilize the airflow over the outboard portion of the wing. Subsequently, the F5D-1 was modified by mounting wooden extensions to the wing leading edge to model the Ogee design (fig. 32). Stew Rolls was the lead engineer on the project, and Fred Drinkwater performed the flight tests. Even under aggressive maneuvering, a stable vortex configuration was observed, relieving concerns about abrupt disturbances to the aircraft from asymmetric bursting of the vortices.[12] As noted in reference 36, the pilots were able to decrease the approach speed of the aircraft by 10 knots, reflecting the improved flight characteristics of the Ogee over the more conventional planform of the F5D-1. Data were provided to the Anglo-French team, which was in the process of designing the Concorde, giving it assurance that the Ogee planform was suitable for the aircraft.

In a postlude to this area of research, in the 1970s flight tests were performed to determine the response of aircraft when encountering trailing vortices in the wake of wide-body transport aircraft. This activity was carried out as part of a joint program

Figure 32
Douglas F5D-1 Skylancer with Ogee wing planform.

---

[10] Stew Rolls 1998: personal communication.
[11] Bill Harper 1998: personal communication.
[12] Fred Drinkwater 1998: personal communication.

Figure 33

Lear 23 in formation with Dryden B-747 and T-37.

with the FAA that involved Ames, Langley, and the Flight Research Center to help the FAA determine if it was feasible to reduce aircraft separation during the terminal-area approach as a means of increasing airport capacity. The objectives of the flight program were to document the magnitude of wake-vortex upsets for different pairs of generating and trailing aircraft, to investigate different ideas for reducing the vortex strength, and to develop methods to predict the magnitude of a trailing aircraft's upset in the wake of a lead aircraft.[13] Along with other aircraft from the Flight Research Center, the Ames Lear 23 (fig. 33) was flown into the wakes of a Boeing 747 and Lockheed C-5A to obtain these measurements. Robert (Bob) Jacobsen was the project engineer and Fred Drinkwater and Glen Stinnett carried out much of the flying (ref. 37). Richard Kurkowski had performed earlier tests of the wake of a Boeing 727. The program included tests in the 40- by 80-foot wind tunnel to obtain measurements of wake-vortex size and circulation using a newly developed laser velocimeter in order to see how aircraft configuration changes affected vortex strength. Flight tests employed the velocimeter for the same purpose. Experiments were also carried out on the Six-Degree-of-Freedom simulator by Robert Sammonds to develop valid simulation techniques based on flight experience and to extend the flight results. Bruce Tinling (ref. 38) and Barbara Short and Bob Jacobsen (ref. 39) generalized the flight experience using methods developed from the wind tunnel tests to predict the bank-angle upset imposed on a variety of aircraft types when following a heavy transport at different distances. This work contributed to the definition of the FAA's separation criteria for landing behind large heavy aircraft.

Aerodynamic research in flight, with the exception noted in the preceding paragraph, was concluded with the F5D-1 tests in 1961. At that time, high-performance flight research was transferred by NASA headquarters directive to the Flight Research Center at Edwards Air Force Base. The results of Ames' work over two decades led to a better understanding of aerodynamic effects on performance and loads in subsonic, transonic, and supersonic flight. Out of flight exploration of the transonic regime came the realization that there was an aerodynamic continuum from subsonic to supersonic flight with no aerodynamic impediment or "barrier" in the flow. Aerodynamicists were thus encouraged to press for wind tunnel facilities that could explore this flow region more generally.[14] George Cooper received the Octave Chanute and Arthur S. Flemming Awards in 1954 for his numerous contributions as a test pilot in several of these programs.

---

[13] Bob Jacobsen 1998: personal communication.
[14] Bill Harper 1998: personal communication

Flying Qualities, Stability and Control, and Performance Evaluations

As a consequence of the United States involvement in World War II, a number of aircraft were sent to Ames by the military services for the principal purpose of obtaining flying qualities evaluations and stability and control and performance assessments using the expertise of Ames pilots and engineers. Many of these aircraft spent a relatively short time at the laboratory, but a few were modified substantially and then used in more extensive research programs.[15] The individual programs had several objectives, including exploratory evaluation of performance and flying qualities and, in some cases, in-depth investigation of specific problems and evaluations of design modifications. The results of these evaluation programs provided a wealth of data, not only useful as they related to the individual aircraft themselves, but as a source from which the military drew to develop flying qualities specifications for future aircraft.[16] It is not possible to comment on the results of all of these tests, for few formal NACA reports were prepared. Instead, the information was disseminated to the military and the manufacturers through informal memoranda that are lost to the authors. In some cases, the only remaining information about these programs comes from Don Heinle's brief notes and from personal recollections of individuals who took part. The following discussion provides an indication of the variety of investigations that were conducted during those years. Table 4 lists the many aircraft that were part of this activity. Photographs that are available for several of the World War II aircraft appear in figures 34–53.

The aircraft at Ames that were heavily instrumented for flying qualities evaluations and stability and control measurements were the A-20A Havoc, B-25D Mitchell, A-35A Vengeance, B-26B Marauder, BT-13B Valiant, PV-1 Ventura, XP-75A-1 Eagle, XP-47M-1 Thunderbolt, P-51F Mustang, F4U-4 Corsair, P-61A-5 Black Widow, SBD-1P Dauntless, and the XP-70 (the first American night fighter, converted from the A-20 Havoc). The B-25 was also used for engine-out control tests. Seth Anderson, William Turner, Thomas Keller, Richard Spahr, and Robert Reynolds all participated in these tests. Flying qualities evaluations were also carried out on the XF7F-1 Tigercat, a short-coupled aircraft with two large radial engines. It underwent modification to enlarge the vertical stabilizer based on development tests in the 40- by 80-foot wind tunnel. George Cooper conducted critical tests of engine-out wave-offs to assess the effects of the larger fin on directional stability and control. Tests showed flying qualities to be satisfactory with the exception that roll performance did not meet the military's specifications (ref. 40). Flying qualities tests on the A-26B Invader led to modifications to the longitudinal and lateral control systems to reduce maneuver control forces based on results of tests obtained in the 40- by 80-foot wind tunnel. Flight evaluations confirmed that the control forces had been reduced to an acceptable level. The FM-2 Wildcat was used for evaluation of carrier landings.

Two OS2U-2 Kingfisher, a Navy scout aircraft, underwent testing and modification to improve performance and longitudinal control. Modifications included the installation of full-span maneuver flaps (Zap Flaps), spoiler-type ailerons, and the

Figure 34
Douglas SBD-1P Dauntless.

Figure 35
Vought-Sikorsky OS2U-2 Kingfisher.

Figure 36
Brewster F2A-3 Buffalo.

---

[15] Harry Goett and Bill Harper 1998: personal communication.
[16] Seth Anderson 1998: personal communication.

Figure 37
OS2U-2 Kingfisher.

Figure 38
Douglas A-20A Havoc.

Figure 39
North American B-25D Mitchell.

TABLE 4. AIRCRAFT USED FOR FLYING QUALITIES, STABILITY AND CONTROL, AND PERFORMANCE EVALUATIONS		
Aircraft Name	Arrival or First Flight Date	Departure Date
SBD-1P	October 24, 1941	February 21, 1942
OS2U-2 (Bu. No. 2189)	March 9, 1942	1944
F2A-3 (Bu. No. 01516)	May 21, 1942	1943
OS2U-2 (Bu. No. 3075)	February 6, 1943	May 24, 1943
A-20A (AAC39-726)	March 10, 1943	May 31, 1943
B-25D (AAF41-29983)	March 26, 1943	1943
A-35A (AAF41-31174)	June 16, 1943	1943
B-26B (AAF41-31702)	September 27, 1943	October 27, 1943
PV-1 (Bu. No. 48871)	January 6, 1944	1944
P-51 D (AAF44-13257 NACA 108)	March 28, 1944	June 2, 1944
P-61A-5 (AAF42-5572)	April 20, 1944	November 16, 1944
BT-13B (AAF42-90461)	April 22, 1944	October 22, 1945
BT-13B (AAF42-89854)	April 25, 1944	May 17, 1945
BTD-1 (Bu. No. 04968)	July 28, 1944	June 30, 1947
XP-70 (AAC39-735)	August 19, 1944	August 21, 1944
XF7F-1 (Bu. No. 03550)	September 2, 1944	June 19, 1948
XP-75A-1 (AAF44-44550)	November 22, 1944	February 7, 1946
XP-47M-1 (AAF42-27385)	February 2, 1945	August 8, 1945
FR-1 (Bu. No. 39650, 39656, 39657, 39659, 39660, 39665)	February 17, 1945	June 1, 1947
FM-2 (Bu. No. 73700)	March 13, 1945	February 25, 1946
P-51F (AAF43-43332)	April 30, 1945	November 6, 1947
SB2D-1 (Bu. No. 04971)	May 18, 1945	October 31, 1947
A-26B (AAF43-4307)	July 4, 1945	January, 1951
F4U-4 (Bu. No. 97028)	August 21, 1945	April 30, 1947
K-21 Airship	circa 1945	circa 1945
P-80A (AAF44-85099)	January 30, 1946	April 10, 1950
F7F-3 (Bu. No. 80372, 80521, 80526, 80546)	February 8, 1946	December 9, 1949
XBT2D-1 (Bu. No. 09086)	March 11, 1946	September 4, 1947
F8F-1 (Bu. No. 94819)	April 2, 1946	June 1, 1953
L-4 Cub (NC 254)	October 5, 1948	October 25, 1948
F-84C (AAF47-1530)	October 19, 1948	October 29, 1948
F-86A (AAF48-291 NACA 116, 47-609 NACA 135)	August 29, 1949 / April 10, 1950	January 11, 1960 / March 15, 1956
XR60-1 (85163)	November 21, 1949	May 18, 1950
F6U-1 (Bu. No. 122483, 122491 NACA 138)	July 23, 1950 / August 24, 1950	August 5, 1953 / October 5, 1953
YF-86D (AF 50-577 NACA 149)	June 26, 1952	February 15, 1960
F-86F (AF 52-4535 NASA 228)	October 10, 1953	September 13, 1965
F-84F-5-RE (AF 51-1364 NACA 155)	October 31, 1953	March 7, 1957
F-94C-1 (AF 50-956 NACA 156)	July 29, 1954	November 18, 1958
FJ-3 (Bu. No. 135800)	September 3, 1954	April 30, 1956
F9F-4 (Bu. No. 121156)	October 21, 1954	August 10, 1955
F9F-6 (Bu. No. 128138)	May 9, 1955	August 3, 1955
F7U-3 (Bu. No. 129656)	June 10, 1955	October 4, 1955
F4D-1 (Bu. No. 134759)	April 4, 1956	October 16, 1959

TABLE 4. AIRCRAFT USED FOR FLYING QUALITIES, STABILITY AND CONTROL, AND PERFORMANCE EVALUATIONS (Concluded)		
Aircraft Name	**Arrival or First Flight Date**	**Departure Date**
F-100C (AF 54-1964) T-ALCS	March 22, 1957	February 15, 1960
F5D-1 (Bu. No. 142350b NASA 213)	August 20, 1957	June 15, 1961
T-33A-5 (AF 49-920A NASA 720)	November 27, 1957	September 15, 1965
F8U-3 (Bu. No. 147085)	June 18, 1959	August 15, 1960
B367-80 (N70700)	April 1967	July 1967
	May 1968	August 1968

VARICAM, or variable camber tail. In the latter case, the percentage of horizontal stabilizer chord dedicated to the elevator was increased and sectioned into two deflecting segments, thus effectively varying the camber of the entire stabilizer. Flap effectiveness was nonlinear and was observed to fall off with flap deflection. Roll control was not satisfactory, having poor forces and feel characteristics (ref. 41). The VARICAM did prove to be satisfactory and increased pitch-control effectiveness somewhat (ref. 42).

The lift-drag polar and static stability margins of the F2A-3 Buffalo were established. The Buffalo's wing was subjected to a static torsion test, and wing torsion was measured in flight.

Variations on engine tilt, control force characteristics, wing dihedral, and aileron shape were evaluated on several FR-1 Fireball aircraft, in a collaborative effort with the testing done in the 40- by 80-foot wind tunnel to rectify several stability and control deficiencies. Some of this testing motivated the development of variable stability aircraft, as noted in the next section of the report. As can be seen in figure 49, the FR-1 had a reciprocating engine in the nose and a small GE I-16 jet engine in the rear. It had been developed by the Navy late in World War II as a fighter that combined conventional propeller-driven and jet propulsion. FR-1 test pilots would occasionally shut down the piston engine and then fly alongside another aircraft, leaving its pilot to wonder how the FR-1 could maintain position with an inoperative engine.[17]

Flight tests with the SB2D-1 confirmed the results of extensive wind tunnel tests in the 7- by 10-foot and 40- by 80-foot wind tunnels, which predicted poor performance because of high drag and poor roll control at low speed.

And, to round out the ensemble of unusual craft, the Navy K-21, a non-rigid airship, was tested to determine if it could be flown acceptably through a servo control system that employed a B-29 autopilot formation control stick.[18]

As a consequence of all of Ames' wartime testing, the laboratory developed a reputation for its work in determining the flying qualities of a wide range of aircraft; this work continued after the war with a number of other projects. Again, many of these

Figure 40
Vultee A-35A Vengeance.

Figure 41
Martin B-26B Marauder.

Figure 42
Lockheed PV-1 Ventura.

---

[17] George Cooper 1998: personal communication.
[18] Ron Gerdes 1998: personal communication.

Figure 43
P-51D Mustang.

Figure 44
Northrop P-61A-5 Black Widow.

Figure 45
Douglas BTD-1 Destroyer.

Figure 46
Grumman XF7F-1 Tigercat.

Figure 47
General Motors XP-75A-1 Eagle.

Figure 48
Republic XP-47M-1 Thunderbolt.

Figure 49
Ryan FR-1 Fireball.

Figure 50
Grumman FM-2 Wildcat.

Figure 51
Douglas A-26B Invader.

were of a short duration with the objective to obtain data concerning problems with a specific aircraft. The collection of post-war aircraft is shown in figures 54–63.

Stability and control and flying qualities evaluations were carried out on a number of high-performance propeller and jet aircraft, including the XBT2D-1 (Skyraider prototype), F6U-1 Pirate, F7F-3 Tigercat, F8F-1 Bearcat, F-84C Thunderjet, F-84F-5-RE Thunderstreak, and the F5D-1 Skylancer. The F8F-1 was also used to examine buffet, including tests with the propeller feathered and engine shut down to permit the aerodynamic contribution to be identified. The F-86A Sabre underwent stall and spin testing. The prototype of the Boeing 707 commercial jet transport, the 367-80, was used for developing flying qualities criteria pertaining to large transport aircraft designs. That program was run by Hervey Quigley.

Several specialized tests were performed for a variety of purposes. In this category, the F-100C T-ALCS demonstrated a normal acceleration-command control system. The T-33A-5 Shooting Star performed zero-g flights and was used for pilot physiological studies. Tail-load tests were carried out on the XR60-1 Constitution. Finally, the L-4 Cub was used in evaluations of a castering landing gear for takeoff with 90 degree crosswind. The latter was one of Bill McAvoy's final test programs.

Two studies were carried out on the B-47 Stratojet, one concerning measurement and prediction of response characteristics of a flexible airplane to elevator control (ref. 43), the second on experimental and predicted longitudinal and lateral-directional response characteristics of a swept-wing airplane (ref. 44). The results provided an indication of the detail required in the analytical models to adequately predict the aeroelastic behavior of the airplane. These studies were carried out by Henry Cole and Stuart Brown and were performed jointly with the NASA High Speed Flight Station. All flights for both programs were conducted at Edwards Air Force Base.

These focused test programs served a useful purpose for the manufacturers and the military in resolving problems with the various designs. Along with them, more enduring efforts were carried out at Ames that had a broader impact on the technology. In one case, tests in the 7- by 10-foot wind tunnel were used to develop predictions of flying qualities, particularly concerning the influence of propeller slipstream effects on stability and control. Harry Goett, Roy Jackson, and Steve Belsley published the summary report of this work (ref. 45), which instigated flight tests with a number of aircraft, most extensively with the Navy's twin-engine patrol aircraft, the PV-1, and lent credibility to the prediction methods (refs. 46 and 47). The flights showed that the wind tunnel results anticipated the unsatisfactory longitudinal characteristics attributed to high control forces in maneuvers and landings. Power effects were confirmed to be critical contributions. The high aileron and rudder forces, which adversely affected roll and engine-out directional control, were also substantiated. Examples of results from other programs which worked their way into the military's flying qualities design specification appear in references 48–57.

In 1947, the Octave Chanute award was given to Larry Clousing in recognition of his contributions to the flying qualities evaluations of a number of the early aircraft and for his work in aerodynamics experiments.

Figure 52
Chance Vought F4U-4 Corsair.

Figure 53
K-21 Airship.

Figure 54
P-80A.

Figure 55
Douglas XBT2D-1 (Skyraider prototype).

Figure 56
Grumman F8F-1 Bearcat.

Figure 57
Taylorcraft L-4 Cub with Seth Anderson.

Figure 58
North American F-86A Sabre.

Figure 59
Lockheed XR60-1 Constitution.

Figure 60
Vought F6U-1 Pirate.

Figure 61
Republic F-84F-5-RE Thunderstreak.

Figure 62
Vought F7U-3 Cutlass.

Figure 63
Boeing 367-80 (prototype for the 707 jet transport).

As a consequence of its long involvement in flying qualities assessments of a wide variety of aircraft, Ames was called upon for specific projects that were of concern to the military services. One particular program stands out in this regard. In the mid-1950s, the Navy was intent on establishing the influence of flying qualities on the minimum acceptable approach speed for landing on an aircraft carrier, and turned to Ames to carry out the program. That effort, led by Maurie White, involved the evaluation of 10 aircraft in 41 different configurations. The Navy sent different aircraft every 2 months to Ames to be instrumented and flown in that program,[19] including the F4D-1, F7U-3, and F9F-6; in addition, the FJ-3 and F9F-4, which were also involved in boundary-layer control research, were also used. Along with these five, Ames flew five Air Force aircraft to broaden the sample, including the F-84F-5-RE, the F-86 E and F, the F-94C, and the F-100A. Configuration variations included flap type and setting, wing leading-edge configuration and flow-control devices, and boundary-layer control systems. In that experiment, extensive pilot opinion data were obtained concerning the stability and control characteristics that influenced the acceptable approach speed. The subsequent report (ref. 58) included comparisons with existing approach-speed selection criteria.

Most of the flights were conducted by George Cooper, Bob Innis, and Fred Drinkwater and took place at the remote test site at the Crows Landing Naval Auxiliary Landing Field in the central valley area east of Moffett Field. Figure 64 shows the mirror landing aid adjacent to the Crows Landing runway that was used for approach guidance. This program was one of the earliest in which ground-based simulation began to play a complementary role with flight test in assessing flying qualities. White and Drinkwater carried out a study of the effects on selection of approach speed using the most rudimentary device. The representation of the external visual scene was provided by a cathode ray tube, which presented an artificial horizon and an outline of the carrier deck. A voltmeter served as the airspeed indicator. Throttle and center stick controls were provided, the latter with fixed spring restraints, with the pilot sitting in front of the lot on a swiveling stool. Everything was linked through an analog computer that performed the computation of the aircraft's dynamic response. Still, the results that were obtained helped to generalize the results that were obtained in flight, and the two together gave a clear indication of the best choice for the desired approach speed. At the culmination of these activities, Drinkwater, Cooper, and Innis all presented their views on the subject in references 59 and 60, with Innis introducing a STOL transport, the YC-134A, to the collection of fighters surveyed. These three men, along with the other members of the research pilot staff as they appeared in 1955, are shown in figure 65.

Another extensive flying qualities investigation that involved an early simulator and a number of airplanes was Brent Creer's study of lateral control requirements. This was carried out on the F6F, F-86, F4D, T-37, F-100, and the P-80A. It also used the pitch-roll chair simulator, a new device with two rotational degrees of freedom, dubbed the NE2 for "any two" axes of motion. This study screened several candidate

Figure 64
Mirror landing aid at Crows Landing, Calif.

---

[19] George Cooper 1998: personal communication.

Figure 65

Flight Operations Branch circa 1955. From left to right: Bob Innis, Don Heinle, Larry Clousing, Bill McAvoy, Fred Drinkwater, George Cooper.

flying quality parameters and showed where motion simulation proved of value in the process (ref. 61).

In concluding this section, it is appropriate to highlight what may be the most important contribution of the flying qualities evaluation programs and experiments conducted on the variable stability aircraft at Ames. This, of course, was George Cooper's standardized system for rating an aircraft's flying qualities. Cooper developed his rating system over several years as a result of the need to quantify the pilot's judgment of an aircraft's handling in a fashion that could be used in the stability and control design process. This came about because of his perception of the value that such a system would have, and because of the encouragement of his colleagues in this country and in England who were familiar with his initial attempts. Characteristically, Harry Goett spurred Cooper on in pursuit of this objective.

Cooper's approach forced a specific definition of the pilot's task and of its performance standards. Further, it accounted for the demands the aircraft placed on the pilot in accomplishing a given task to some specified degree of precision. The Cooper Pilot Opinion Rating Scale was initially published in 1957 (ref. 62). After several years of experience gained in its application to many flight and simulator experiments and through its use by the military services and aircraft industry, it was subsequently modified in collaboration with Robert (Bob) Harper of the Cornell Aeronautical Laboratory and became the Cooper-Harper Handling Qualities Rating Scale (fig. 66) in 1969 (ref. 63). This rating scale has been one of the enduring contributions of flying qualities research at Ames over the past 40 years; the scale remains as the standard way of measuring flying qualities to this day. In recognition of his many contributions to aviation safety, Cooper received the Adm. Luis de Florez Flight Safety Award in 1966 and the Richard Hansford Burroughs, Jr., Test Pilot Award in 1971. After he retired, both he and Bob Harper were selected by the American Institute of Aeronautics and Astronautics to reprise the Cooper-Harper Rating Scale in the 1984 Wright Brothers Lectureship in Aeronautics.

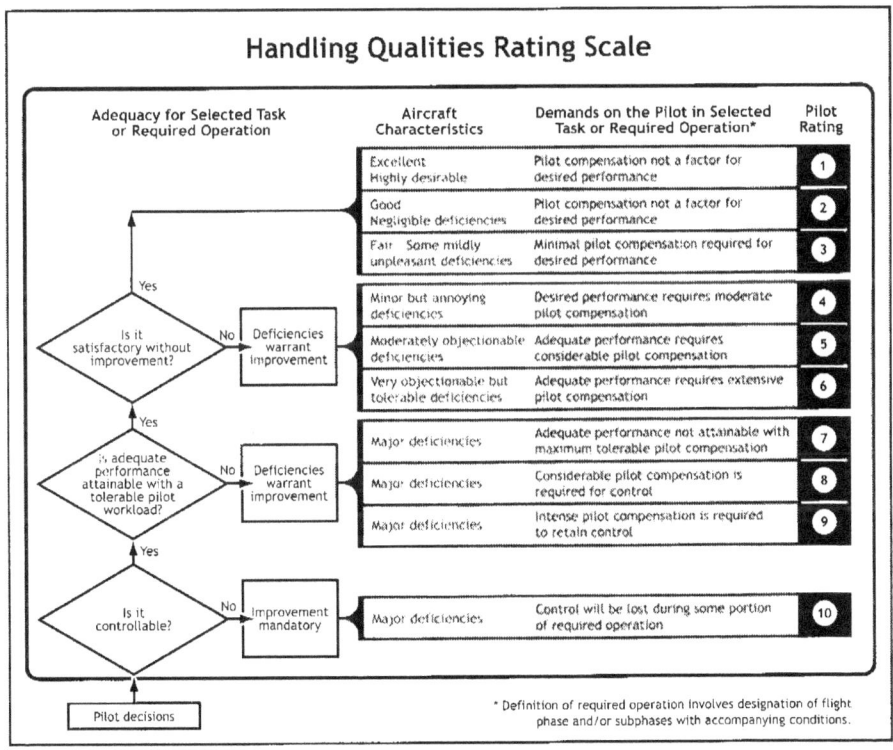

Figure 66
Cooper-Harper Handling Qualities Rating Scale.

Variable
Stability Aircraft

In 1948, an F6F-3 Hellcat was modified by Ames engineers to become the world's first variable stability aircraft. The genesis of this idea followed an investigation (noted previously) into the desired wing dihedral for the Ryan FR-1 Fireball. Three of these aircraft were built, each with a different dihedral angle, to narrow the final design option. This cumbersome and time-consuming approach to a solution inspired William (Bill) Kauffman to develop the concept of a variable stability aircraft.[20] Kauffman had the idea that the basic flight characteristics of such an aircraft could be altered by a stability-augmentation system, so that a wide range of static and dynamic characteristics, representing the flying qualities of a different aircraft, could be safely simulated and evaluated in flight. In the initial design of the F6F-3, the variable stability system altered the effective wing dihedral by deflecting the aircraft's ailerons in response to sideslip. The cockpit control linkage was modified so that the pilot's control stick would not move in response to the aileron deflections commanded by the variable stability system, yet it allowed the pilot to control the roll axis conventionally. Later, modifications to the variable stability system allowed it to command rudder deflections in response to roll rate, yaw rate, and sideslip angle, while roll rate and sideslip were fed into the existing aileron deflection system. Contributions to the development of the variable stability system, along with those of Bill Kauffman, came from G. Allan Smith, an expert in servomechanism design. Reference 64 describes the modifications that were made to develop this variable stability airplane.

Thus configured, the F6F-3 was used for generalized studies of lateral-directional flying qualities criteria and as an in-flight simulator for new aircraft under development. In the latter case, the industry test pilots were able to experience the flying qualities of a new design before its first flight. In a prominent example of the results of these programs, Lockheed was led to incorporate negative dihedral into the design of the F-104 Starfighter after observing the effects on lateral-directional flying qualities of dihedral ranging from positive to negative. This design feature, quite startling and unconventional for most aircraft at that time, was based on a compromise between the expected Dutch roll oscillatory characteristics and the roll response to rudder deflection. Throughout the time the F6F was used as a variable stability aircraft, Rudy Van Dyke, Don Heinle, and Fred Drinkwater were the principal test pilots involved in the programs. Individual members of the Flight Research and Flight Operations Branches, many of whom were associated with the F6F, are shown in figure 67.

Following the development of the F6F, and as high-performance swept-wing jet aircraft came to the fore, evolution of the variable stability concept came about on a series of F-86 aircraft and eventually on an F-100C. To develop lateral-directional flying qualities requirements for the high-performance aircraft of that time, an F-86A and later an F-86E were modified for variable stability. The F-86A's variable-stability system was limited to the yaw axis, whereas that of the F-86E included the roll and yaw axes. Another aircraft, a YF-86D, was used to evaluate longitudinal control system features. Variable stability in this case consisted of changes in control system

---

[20] Harry Goett 1998: personal communication

Figure 67

Flight research personnel with Grumman F6F-3 Hellcat variable stability airplane circa 1950. Front row: Howard Matthews, Allen Knox, Brent Creer, Gus Brunner, Lou Smaus, Bill Kauffman, Marv Shinbrot, Bill McAvoy, Lee Winograd, Walt McNeill, Marv Abramovitz, Merle Waugh, Ed Ernst. Second row: Howard Ziff, Harvey Brown, Carl Tusch (Air Force Liaison), Unidentified, Rose Teresi, Judy Sadoff, Carolyn Hofstetter, Rita Bird, Marjorie Barr, Bobbie Rodenborn, Zada Longanecker. Third row: Steve Belsley, Seth Anderson, Harry Greenberg, Smokey Patton, Carl Hanson, Ben Mayo, Dick Bray, George Rathert, Stew Rolls, Howard Turner, Don Heinle, Mary Lou Reed, Larry Clousing. Norm McFadden.

Figure 68

F6F-3 variable stability airplane.

feel, such as sensitivity and breakout force, in addition to control system dynamic response. One of the latter investigations of the F-86s involved the determination of minimum allowable stability requirements, a subject of interest subsequently in the design of computer-controlled aircraft. Several engineers led the various programs on these aircraft; among them were Charles Liddell, Walter (Walt) McNeill, Brent Creer, Norman McFadden, Richard (Dick) Vomaske, and Frank Pauli. Results representative of investigations into the desired dihedral effect, lateral damping and oscillatory response characteristics, lateral-directional coupling, longitudinal stability, and longitudinal control system characteristics are contained in references 65–73. All these aircraft that were configured as variable stability vehicles are noted in table 5 and shown in figures 68–71.

TABLE 5. AIRCRAFT MODIFIED FOR VARIABLE STABILITY USE		
Aircraft Name	**Arrival or First Flight Date**	**Departure Date**
F6F-3 (Bu. No. 42874 NACA 158)	June 22, 1945	September 9, 1960
F-86A (AAF47-609 NACA 135)	April 10, 1950	March 15, 1956
YF-86D (AF 50-577 NACA 149)	June 26, 1952	February 15, 1960
F-86E (AF 50-606A NACA 157)	June 30, 1955	November, 1959
F-100C (AF 53-1709A NASA 703)	September 4, 1956	November 2, 1960
	March 11, 1964	May 21, 1972
X-14A/B (AF 56-4022 NASA 234, NASA 704)	October 2, 1959	May 29, 1981
CH-47B (USA 66-19138 NASA 737)	August 14, 1979	September 20, 1989

Figure 69
F-86A variable stability airplane.

During the years that the F6F and F-86s flew as variable stability test beds, a variety of new aircraft designs were simulated in order to investigate their flying qualities for a range of piloting tasks. The new designs included the D-558-II, XF-10F, X-1, B-58, XF-104, XF8U-1, F9F-9, XT-37, B-57D, T-38, and the P6M. They ran the gamut from high-performance fighters to bombers, illustrating the breadth of capability in Ames' stable of variable stability aircraft at that time. Bill Kauffman's genius in the conception and development of these aircraft was widely recognized and appreciated by his peers. His original variable stability system design was granted a U.S. patent in 1955. For his extensive work over the years on in-flight simulation, he was presented the Arthur S. Flemming Award in 1955 as one of 10 outstanding young men in the federal service.

Figure 70
YF-86D variable stability airplane.

The last high-performance fighter developed with the variable stability capability was the F-100C, the first three-axis (pitch, roll, and yaw) variable stability aircraft at Ames. A team led by John V. Foster (fig. 72) undertook the system design for this aircraft, which is described in reference 74. After serving as an in-flight simulator for one new, highly advanced aircraft in 1960, it was transferred to the Flight Research Center in response to the headquarters directive at that time. It was later returned to Ames, and Walt McNeill then led a program to evaluate direct-lift control as an aid to air-to-air refueling. Jack Ratcliff was in charge of the effort to modify the variable stability system for direct-lift control. Bob Innis and Ron Gerdes both flew as subject pilots and executed actual plug-ins to the drogue on a KC-135 tanker. The program demonstrated the improvement in precision the pilot could achieve in control of the probe engagement with the drogue when using direct-lift control (ref. 75).

The X-14A and B and the CH-47B were also very productive variable stability aircraft that were used to develop V/STOL and rotary-wing aircraft flying qualities criteria. These aircraft are listed in the V/STOL and rotorcraft categories, but their research systems trace their lineage to the original variable stability aircraft.

Figure 71
F-86E variable stability airplane.

Figure 72
North American F-100C Super Sabre variable stability airplane team. From left to right: Don Heinle, Mel Sadoff, Dick Bray, Walt McNeill, G. Allan Smith, Jack Ratcliff, John Foster, Jim Swain, Howard Clark, Don Olson, Dan Hegarty, Gil Parra, Eric Johnson, Fred Drinkwater.

Gunsight Tracking and Guidance and Control Displays

The control system expertise that Ames engineers were beginning to acquire went beyond variable stability aircraft. Using the new capabilities available with electronic systems, efforts were expanded into the areas of guidance, control, and displays, and a number of aircraft were adapted for those purposes (table 6). A particular demand for this technology at the time involved precision tracking of a target aircraft. In order to understand the contributions of the aircraft's response to precision tracking, a series of flight tests was performed in which the tracking performance of two straight-wing fighters, the P-51H and the F8F-1, was compared with that of two swept-wing candidates, the F-86A and F-86E. Each aircraft used a fixed gunsight. George Rathert and Burnett Gadeberg were the principal engineers responsible for these tests; the results were reported in reference 76.

Sometime later, a lead-computing sight was evaluated in the F-86D in another development effort and test carried out by Gadeberg and Rathert. It was observed that the pilots were able to compensate for a wide range of stability and control characteristics without any variation in tracking performance when using this sight (refs. 77 and 78). Until this time, another aircraft had acted as the "target" for these tracking tests. As a way of cutting costs and improving flight safety, Ames engineers developed a method of simulating the maneuvering target aircraft using equipment on the tracking aircraft itself. In this case, the pilot tracked a spot of light projected onto the windscreen as described in reference 79 and shown in figure 73. The spot

TABLE 6. AIRCRAFT MODIFIED WITH GUNSIGHT TRACKING AND GUIDANCE AND CONTROL DISPLAYS		
Aircraft Name	Arrival or First Flight Date	Departure Date
F8F-1 (Bu. No. 94819)	April 2, 1946	June 1, 1953
P-51H (AAF44-64415 NACA 130)	December 18, 1946	April, 1961
R4D-6 (Bu. No. 99827 NACA 18, NASA 701)	December 14, 1948	September 9, 1965
SB2C-5 (Bu. No. 83135 NACA 147)	December 18, 1948	June, 1955
F-86A (AAF48-291 NACA 116)	August 29, 1949	January 11, 1960
F6F-5 (Bu. No. 79669 NACA 208)	June 19, 1950	September 9, 1960
F-86E (AF 50-580)	April 8, 1952	April 18, 1952
F-86D (AF 51-5986)	June 12, 1953	November 7, 1957
F-86F (AF 52-4535 NASA 228)	October 10, 1953	September 13, 1965
TV-1 (P-80C Bu. No. 33868 NACA 206)	October 12, 1953	February, 1960
F-84F (AF 51-1346)	March 1, 1954	March 11, 1954
F9F-8 (Bu. No. 131086)	January 6, 1955	February 7, 1955
F-86D-5 (AF 50-509A)	January 6, 1955	April 3, 1956
F-86D-5 (AF 53-787 NACA 216)	March 17, 1955	February 1, 1960
F-102A (AF 56-1304)	April 10, 1957	Unknown
F-102A (AF 56-1358)	December 23, 1957	March 21, 1960
F-106A (AF 57-235)	September 4, 1958	December 14, 1959
CV-340 (NASA 707)	May 21, 1963	September 3, 1976
Cessna 402B (NASA 719)	June 5, 1975	May 6, 1982
Boeing 727	1981	1981
Beech 200 (NASA 701)	August 5, 1983	October 3, 1997

would move in the same manner as the image of a maneuvering target; the main drawback of this simulation was that the tracking pilot could not perceive the "attitude" of the target, and could not anticipate the next maneuver. Brian Doolin and G. Allan Smith were in charge of this part of the program, and Fred Drinkwater was the pilot. This target simulator was eventually adapted for simulations of the guidance of a radio-controlled missile, the Bullpup, with Joseph Douvillier and John V. Foster responsible for its development (ref. 80). The TV-1 (a Navy trainer version of the P-80) aircraft served as the test bed for this simulator, and the simulator proved so successful that it was eventually used to train Navy pilots in Bullpup operations. Fred Drinkwater again was the principal pilot for these evaluations. As a consequence of this success, the team developed a target simulator for the Air Force's E-4 radar scope presentation fire-control system.[21] Foster led the effort, including the installation on the F-86D. Flight results showed that the target simulator duplicated the attack phase for an actual airborne target, and further that it might be useful for pilot training as well (ref. 81).

Control systems were eventually developed to allow remote piloting of one aircraft from another.[22] This was first demonstrated when an SB2C-5 was remotely flown

Figure 73
Target tracking display.

[21] John V. Foster 1998: personal communication.
[22] Howard Turner 1998: personal communication.

from an F6F-5, in a project led by Howard Turner and John White. The mother aircraft (the F6F-5) was flown by Rudy Van Dyke. The beep control proved generally satisfactory for remote-control flight tests, including takeoff and landing, except for rollout after touchdown in crosswinds (ref. 82). Although the Navy lost interest in this kind of remotely flown vehicle, the control system installed in the SB2C-5 was adapted to another test. In this case, the aircraft was flown as a sort of "radar-controlled" interceptor; the pilot visually tracked the target with a periscope, and the aircraft would in turn respond to the motions of the periscope to track the target. This system was tested in simulation before flight. Howard Turner, William Triplett, and John White carried out the experiment (ref. 83).

The F-84F aircraft was lent to Ames for a brief time for fixed-sight tracking tests. Although documentation is sketchy, it appears that the F9F-8 Cougar aircraft was at Ames to evaluate the application of an A-1 sight to a "toss-bombing" technique. The F-102A and F-106A tests that involved Ames consisted of evaluations of the fire-control and auto-maneuvering systems. The fire-control system used in the F-106A was designated MA-1. One of the F-102A aircraft flew with an adaptive control normal acceleration command system that was able to maintain consistently satisfactory response characteristics from landing approach to low supersonic speeds at altitude (ref. 84).

Research on artificial vision for landing marked the beginning of display work at Ames. This project involved the evaluation of a television display of the forward scene in the R4D-6 with variations in field of view. The experiment was carried out by Bernard Kibort and Fred Drinkwater (ref. 85). (Similar conceptual ideas are being explored currently for application to a next-generation High Speed Civil Transport.) Years later (1981), in a program of major significance, a head-up guidance display was demonstrated at Ames in a Boeing 727-100 airplane operated by the FAA. This research was motivated by a series of wind-shear-induced landing accidents in the mid-1970s and became a key element in a joint NASA-FAA investigation of the use of head-up displays for landing approach.

The display concept, conceived and developed by Richard (Dick) Bray (ref. 86), consisted of a flightpath-centered, pursuit-tracking presentation in which the primary controlled element represented the direction of flight of the aircraft. The aircraft was flown by directing the flightpath symbol at outside references such as the intended touchdown point or, for instrument flight operations, at appropriate guidance elements in the display. An example of the HUD image during a visual approach can be seen through the windscreen of the B-727 in figure 74. This concept underwent extensive development by Bray in the flight simulators at Ames and proved to be successful, in part because of the availability of inertial measurements that were sufficiently accurate for the pilot to directly observe and control the flightpath. Earlier HUD tests, carried out on one of the Center's C-8A Buffaloes, had been unsuccessful because the attitude sensors were not sufficiently accurate and because it

Figure 74
Head-up display mounted in the FAA Boeing 727.

was necessary to rely on angle-of-attack sensors to derive flightpath information. The display has subsequently been developed by industry for application to commercial transports and has been certificated by the FAA for operation to low-visibility minimums. Federal Express adopted the display for its cargo operations, and several airlines, including Alaska, Southwest, Delta, and United, now have or are preparing to put this display into service. It has been developed further in simulation and successfully demonstrated in flight experiments for V/STOL fighters and transports and for rotorcraft by Vernon (Vern) Merrick and Charles (Charlie) Hynes. Gordon Hardy made significant contributions to this work, both from his background as a test pilot and as an engineer, including participation in the FAA flight program to certify the display for commercial operation. For his contributions to flight safety, Dick Bray received the Adm. Luis de Florez Air Safety Award in 1984.

Research in inertial navigation was pursued to provide the improved accuracy demanded for operations, particularly in the terminal area, including automatic landings. These efforts grew out of Stanley Schmidt's research in application of the Kalman filter to inertial navigation and to its implementation in airborne computers. The system was initially flown on the Convair 340 in tests conducted at the White Sands Missile Range in a cooperative program between Ames and the NASA Manned Spacecraft Center and the Army Instrumentation Directorate at White Sands. This navigation scheme was of particular interest to the space shuttle program and to White Sands for their tracking systems.[23] Leonard McGee led the program and Gordon Hardy and Glen Stinnett carried out the flight operations. Accuracies sufficient to meet stringent automatic landing requirements were demonstrated and reported in reference 87. Investigations of time-constrained area navigation (4D RNAV) for short takeoff and landing (STOL) aircraft were also pursued on the CV-340. The objective of this research was to reduce the amount of airspace used for STOL operations compared with their conventional transport counterparts. This work validated Heinz Erzberger and Homer Lee's concept for the on-board computation of trajectories that would produce the desired time of arrival at the final approach fix. The 4D RNAV was tested extensively in piloted simulation and then implemented in a digital avionics system (STOLAND) developed for eventual use in an Ames STOL research aircraft. In the flight program, led by Lee with Hardy as the project pilot, time of arrival within 5 seconds of the target was typically achieved. Results also pointed out steps in design of the guidance algorithms that were required to reduce workload for manual control (ref. 88). Erzberger summed up the status of the on-board flightpath generation techniques in reference 89. Later, applications were made to STOL aircraft operations, including flight tests on the QSRA conducted by Charlie Hynes and Erzberger.

In conjunction with this area of research, it should be noted that an integrated digital flight management, guidance and navigation system was developed by an industry team from Honeywell and King Radio under the direction of George Callas and Dallas Denery and demonstrated on a Cessna 402B for general aviation applications (ref. 90). In addition, Charles Jackson conducted tests on DME/DME (distance measuring equipment) navigation and touch-panel displays on this aircraft. Also, in the early 1990s, precision landing guidance research using satellite-based navigation was pursued on the Beech Super King Air by David McNally and Russell Paielli with Rick Simmons as project pilot. The navigation system was interfaced with the aircraft's approach guidance and autopilot system, and coupled approaches were flown to altitudes of 50 feet. Data that quantified the accuracy of the differential global positioning system and landing guidance algorithms were obtained from these flights and, in cooperation with Stanford University, demonstration programs were performed by American and United airlines to assess their readiness for commercial operations (ref. 91).

In order to carry out research in guidance and control, it was necessary at some point to obtain credible mathematical models of the aircraft's response characteristics for use in analyzing the system designs. This required either measures of the aircraft's open-loop response to control inputs or identification of the important aerodynamic force and moment characteristics that determine the aircraft's response. Wind tunnel data could provide reasonable estimates of static stability derivatives such as angle-of-attack or directional stability coefficients, but few facilities were available from which rotational stability derivatives such as pitch, roll, or yaw damping from ground-based

Figure 75
P-51H Mustang.

Figure 76
Douglas R4D-6 Skytrain.

Figure 77
Curtiss SB2C-5 Helldiver.

---

[23] Dallas Denery 1998: personal communication.

Figure 78
F-86E Sabre.

Figure 79
F-86D Sabre.

tests could be obtained. Thus, flight test methods were developed to extract this information during dynamic maneuvers of the aircraft.[24] Early efforts at Ames involved the work of William Triplett in the 1950s to acquire overall measures of the aircraft's frequency response from which transfer functions relating the aircraft's important state variable to its individual controls could be defined (ref. 92). Dallas Denery later made contributions to the identification of the aircraft's individual stability and control characteristics (ref. 93). Rodney Wingrove and Ralph Bach devised methods to extract the best estimate of the aircraft's motions in the presence of uncertainties owing to measurement inaccuracies or external disturbances to the aircraft (ref. 94). The aircraft state estimation methods were ultimately used extensively in aircraft accident and incident analyses by the National Transportation Safety Board, particularly for determining the wind environment in which an aircraft was operating. In recent years, Mark Tischler returned to the use of frequency-response methods to extract transfer functions and stability and control derivatives from flight records and has developed an analysis program (ref. 95) that has found wide use in the aircraft industry and military services.

A number of the aircraft that were used in this research are shown in figures 75–88. Flight research personnel in 1959 are shown in figure 89, and their counterparts in guidance and navigation are shown in figure 90.

[24] Dallas Denery 1998: personal communication

Figure 80
Lockheed TV-1.

Figure 81
Grumman F9F-8 Cougar.

Figure 82
F-86D-5.

Figure 83
Convair F-102A Delta Dagger.

Figure 84
Convair F-106A Delta Dart.

Figure 85
Convair 340.

Figure 86
Cessna 402B.

Figure 87
Boeing 727.

Figure 88
Beech 200.

Figure 89

Flight research personnel circa 1959. Front row: George Rathert, Stu Brown, Norm McFadden, Howard Turner, Gus Brunner, Venia McCloud, Violet Shaw, Kay Rizzi, Yvonne Settle, Genevieve Ziegler, Anita Palmer, Grace Carpenter, Evelyn Olson. Second row: Bill Triplett, Alan Faye, Dick Bray, Seth Anderson, Steve Belsley, Hervey Quigley, Hank Cole, Elwood Stewart, Don Higdon, Maurie White, Dorothea Wilkinson, Dick Vomaske, Stew Rolls, Mel Sadoff, Mary Thompson, Brent Creer. Back Row: Ron Gerdes, Joe Douvillier, John Stewart, Rod Wingrove, Walter McNeill.

Figure 90
Guidance and navigation personnel circa 1969. Front row: Dick Kurkowski, Michele Hilliard, Brent Creer, Grace Webster, Fred Edwards. Second row: Rod Wingrove, Bedford Lampkin, Armando Lopez, Del Watson. Third row: Fred Boltz, Clark White, Gordon Hardy, Don Smith. Back row: Hank Lessing, Dallas Denery, Dick Acken, Bob Coate.

In-Flight Thrust Reversing, Steep Approach Research

The Navy took increased interest in low-speed flight when the introduction of jet aircraft to the aircraft carrier revealed flying qualities problems that had not been experienced with piston-powered aircraft. One concern related to the adverse effect on flightpath control of the jet engine's slow response to the pilot's throttle inputs. Following an early simulation investigation of the selection of approach speeds for landings aboard ship, an in-flight thrust reverser was evaluated as one means to allow pilots to quickly change the longitudinal component of thrust without having to change engine rpm.[25] This concept was investigated in the 40- by 80-foot wind tunnel at Ames in order to determine stability and control influences; flight tests in the F-94C aircraft then followed. The reverser installation on the F-94C can be seen in figure 91. Seth Anderson was the project leader of this work with George Cooper as project pilot. Results demonstrated an improvement in flightpath response to thrust control, an expanded descent flight envelope over a wide range of speeds, and improvements in touchdown precision (ref. 96). A demonstration flight program was arranged for Navy, Air Force, and airline pilots in order to expose them to the use of the thrust reverser in flight. Later, the thrust-reversing concept was applied to the

Figure 91

Thrust reverser on F-94C. From left to right: Air Force Major E. Sommerich, Seth Anderson, Lt. Col. Tavasti, and George Cooper.

---

[25] George Cooper and Seth Anderson 1998: personal communication.

DC-8 commercial transport to achieve the rapid descent capability required for FAA certification.

Terminal-area approach and landing studies continued at Ames with conventional aircraft. The first CV-990 at Ames, which became the airborne science platform, Galileo I, was used for direct lift control (DLC) research in 1968. The aircraft's spoilers were coupled to the flight control system to provide better flightpath response during the approach and landing (ref. 97). Dick Bray was the research leader and Fred Drinkwater and Bob Innis participated as test pilots. This DLC system was demonstrated to engineers and pilots of the airline industry. It was eventually incorporated as part of the Lockheed L-1011 flight control system, and was instrumental in achieving excellent automatic landing performance for that aircraft.

Steep descent testing, including power-off landing approaches and demonstration of minimum lift-to-drag ratio (L/D) landings came out of the interest in the use of low L/D lifting bodies for recovery to landing from space. The question posed to the flight research organization concerned how low an aircraft's L/D could be for the aircraft to still be landed successfully.[26] Flight tests with the JF-104A Starfighter were conducted by Fred Drinkwater, who demonstrated steep approaches that were ultimately used by the space shuttle (ref. 98). These two-segment profiles consisted of a steep upper segment starting around 25,000 feet and aimed at a target a mile short of the runway, followed by a 3-degree path to the touchdown point. These profiles became widely known within the test pilot community as the "Drinkwater Approach." The CV-990 was also used for space shuttle approach and landing studies. Ames guidance and navigation expertise was tapped to develop concepts which would be used by the shuttle in this flight phase. Through Ames' efforts, supported by Sperry's Flight Control Division, a digital navigation, guidance, and autopilot system was installed in the aircraft to test the feasibility of energy-management approach concepts for an unpowered vehicle. Flight tests were carried out in 1972 on the Galileo I and, in 1975, on the second CV-990, Galileo II, by Drinkwater, with technical direction by Fred Edwards and John D. Foster, along with significant input from Gordon Hardy on the pilot's system interface. They developed the circular path geometry following reentry to intercept the final steep straight-in path and provided guidance and energy management for this path with the digital autopilot (ref. 99). Approaches were made with the engines at idle from an altitude of 40,000 feet at speeds of at least 200 knots. Results showed that, with the proper system design, safe approaches and precise landings could be achieved with an unpowered vehicle. This complete system demonstration contributed to the decision to remove air-breathing engines from the final shuttle design.[27]

The second CV-990 was also equipped with a digital navigation/guidance/automatic-landing system to study energy-management landing-approach concepts for commercial jet transports. A "delayed flap" concept that reduced fuel use and approach noise was developed and flown in 1975 (ref. 100). Demonstrations of this system were

---

[26] Fred Drinkwater 1998: personal communication.
[27] Brent Creer 1998: personal communication.

made for airline representatives. Drinkwater flew these tests with technical direction by John Bull, Fred Edwards, and John D. Foster. In a somewhat similar vein, noise-abatement landing approach patterns, which were explored initially on the Boeing 367-80 by Hervey Quigley, were investigated in an extensive program led by Dallas Denery and conducted by an Ames team in collaboration with American Airlines and United Airlines. The purpose of Denery's work was to develop an avionics system that would allow commercial jet transports to perform two-segment landing approaches under instrument flight conditions.[28] The tests, flown in 1971 using an American Boeing 720 and in 1973–74 on a United Boeing 727 and Douglas DC-8 aircraft with crews from the airlines, FAA, and NASA, showed the effectiveness of a two-segment descent profile in substantially reducing noise on the ground under the approach path (ref. 101).

The various aircraft are shown in figures 92–96 and are listed in table 7. Flight operations personnel in 1970 appear in figure 97.

Figure 92
Lockheed F-94C-1 Starfire.

TABLE 7. AIRCRAFT USED FOR IN-FLIGHT THRUST REVERSING AND STEEP APPROACH RESEARCH		
Aircraft Name	Arrival or First Flight Date	Departure Date
F-94C-1 (AF 50-956 NACA 156)	July 29, 1954	November 18, 1958
JF-104A (AF 56-745A)	March 19, 1959	May 6, 1960
CV-990 (NASA 711)	April 2, 1965	April 12, 1973
CV-990 (NASA 712)	December 10, 1973	July 17, 1985

Figure 93
Lockheed JF-104A Starfighter.

Figure 94
Convair 990 Galileo I.

---

[28] Brent Creer 1998: personal communication.

Figure 95

Convair 990 Galileo II.

Figure 96

United Airlines Douglas DC-8.

Figure 97

Flight operations personnel circa 1970. From left to right: Bob Innis, Frank Brasmer, Fred Drinkwater, Dick Gallant, Gordon Hardy, Glen Stinnett, Jean Moorhead, George Cooper, Ron Gerdes, Dan Dugan, Jim Satterwhite.

Boundary Layer Control, STOL, V/STOL Aircraft Research

BOUNDARY LAYER CONTROL RESEARCH

Motivated by the military's interest in reducing landing speeds for their jet fighter aircraft, Ames aerodynamicists began to explore practical ways of controlling the boundary layer of free-stream air on wings, high-lift devices, and control surfaces in the 1950s. Blowing directly over an airfoil increases the circulation of the air over the airfoil, thereby increasing lift and, by energizing the boundary layer, preventing separation of the airflow from the surface. Extensive testing was done in the 40- by 80-foot wind tunnel to explore different approaches for boundary-layer control.[29] This led to flight investigations of a variety of boundary-layer control concepts on the F-86F over the period 1954–57, including suction at the leading edge of the wing, suction at the leading edge of the flap, and blowing over the flap to energize the boundary layer. The blown flap, which eventually saw application on operational aircraft, proved to be more effective than suction at the leading edge of the wing or flap for increasing lift since it increased circulation at the wing whereas suction just enabled the wing to approach its theoretical maximum lift. Results of tests of the three different schemes were reported in references 102–104. Ames continued its flight investigation of boundary-layer control lift augmentation in tests of the FJ-3, F9F-4, and F9F-6 aircraft. The F9F-6 was also used to obtain low-speed lift and drag data during approach.

The F-100A aircraft was used in a test of blown leading-edge and trailing-edge flaps. The system exhibited improved longitudinal stability and reduced the onset of buffet to airspeeds about 35 knots lower than for the conventional aircraft. Contributions to this reduction in speed could be attributed to the addition of short-span trailing-edge flaps (15 knots), blowing on the leading-edge flaps (13 knots), and blowing on the trailing-edge flaps (7 knots).[30] However, control problems began to appear that were not evident at the higher approach speeds (ref. 105). Hervey Quigley led this program with Bob Innis as the principal pilot. The experience gained from all this testing was utilized in designing the blown flap systems used on the F-104 Starfighter and the F-4 Phantom II and in designing the short takeoff and landing (STOL) transport aircraft to come.

Innis teamed with Stew Rolls in another boundary-layer control (BLC) application, this time to a large four-engine jet transport, the Boeing 707 prototype 367-80. Using a system developed through tests with a 30 percent scale model of the aircraft in the 40- by 80-foot wind tunnel, the aircraft was routinely flown at approach speeds 20 to 35 knots below those used in normal operations. Lift and drag data were obtained in flight as a function of engine thrust coefficient. The lift coefficients achieved were not as high as those obtained in the tunnel, a result that was attributed to the effects of model scale on flow as influenced by the leading-edge devices (ref. 106).

The various BLC, as well as STOL and vertical and short takeoff and landing (V/STOL) aircraft, are noted in table 8. The collection of BLC aircraft appears in figures 98–102.

Figure 98
F-86F Sabre.

Figure 99
North American FJ-3 Fury.

Figure 100
Grumman F9F-4 Panther.

---

[29] Bill Harper 1998: personal communication.
[30] Woody Cook 1998: personal communication.

Figure 101
Grumman F9F-6 Cougar.

Figure 102
North American F-100A Super Sabre.

TABLE 8. AIRCRAFT USED FOR BOUNDARY LAYER CONTROL, STOL, AND V/STOL RESEARCH		
Aircraft Name	Arrival or First Flight Date	Departure Date
F-86F (AF 52-4535 NASA 228)	October 10, 1953	September 13, 1965
FJ-3 (Bu. No. 135800)	September 3, 1954	April 30, 1956
F9F-4 (Bu. No. 125156)	October 21, 1954	August 10, 1955
F9F-6 (Bu. No. 128138)	May 9, 1955	August 3, 1955
F-100A (AF 53-1585A NACA 200)	October 2, 1956	February 15, 1960
VZ-3RY (AF 56-6941 NASA 235, NASA 705)	May 20, 1958, August 24, 1959	February 24, 1959 June 20, 1966
YC-134A (AF 54-556 NASA 222)	March 6, 1959	May 31, 1961
XV-3 (AF 54-148)	August 12, 1959	June 9, 1965
X-14 (AF 56-4022)	October 2, 1959	
X-14A (NASA 234)	1960	
X-14B (NASA 704)	1971	May 29, 1981
NC-130B (AF 58-712)	June 30, 1961 February 27, 1963 November 16, 1963	December 20, 1961 May 23, 1963 November 5, 1967
YROE-1(4020, 4021, 4024)	November 16, 1961	1961
CV-340 (NASA 707)	May 21, 1963	September 3, 1976
XV-5B (USA 62-4505 NASA 705)	March 17, 1964	January 30, 1974
B367-80 (N70700)	April 1967 May 1968	July 1967 August 1968
C-8A (USA 63-13686)	June 10, 1967	Delivered to Boeing for modification to AWJSRA
OV-10A (Bu. No. 152881 NASA 718)	April 8, 1968	October 7, 1976
C-8A (AWJSRA NASA 716)	May 1, 1972	September 22, 1981
DHC-6 (NASA 720)	August 7, 1973	September 19, 1979
XV-15 (NASA 702)	March 23, 1978	April 5, 1990
C-8A (USA 63-13687 QSRA NASA 715)	August 3, 1978	November 22, 1993
XV-15 (NASA 703)	October 30, 1980	April 28, 1994
YAV-8B (Bu. No. 158394 NASA 704)	April 11, 1984	November 30, 1995
AV-8C (Bu. No. 158387 NASA 719)	January 21, 1986	February 17, 1995

Figure 103
Stroukoff YC-134A.

STOL RESEARCH

Short takeoff and landing flight research was motivated by the desire of military and civil operators for transport aircraft with short-field operational capability and jet cruise speeds. For Ames, it was a natural extension of the earlier boundary-layer control activity undertaken to achieve low-speed performance. The first STOL flight research at the Center involved two transports that had been developed for the Air Force, the YC-134A and NC-130B (figs. 103 and 104). Both aircraft used boundary-layer control over the flaps to augment lift. For the NC-130B, extensive tests in the 40- by 80-foot wind tunnel had shown the capability of BLC flaps to enable operation at low airspeeds. However, uncertainties still existed about the ability to control the aircraft in this flight regime. To provide the control power required at speeds as low as 60 knots, the NC-130B also used blowing over drooped ailerons, elevator, and rudder. Two engine pods were attached to the wing of the aircraft for the sole purpose of providing bleed air for the entire boundary-layer control system. Results of the NC-130B tests are reported in reference 107. Both aircraft, as modified, proved the low-speed performance anticipated, but they were found by the pilots to suffer from poor lateral-directional control characteristics during the low-speed approach and landing. Control augmentation systems were devised for the NC-130B on a ground-based simulator and then demonstrated in the airplane to improve lateral-directional control at these conditions (ref. 108). Special procedures were developed during the flight tests to reduce the time spent in the low-speed configuration during takeoff and landing to reduce the overall pilot workload.

Figure 104
Lockheed NC-130B.

Ames research with these aircraft brought the team into contact with comparable activities under way in France and Japan. Joint flight programs were undertaken with both countries in order to improve the understanding of a wide variety of STOL configurations and their operations.[31] Two major tests were carried out in France on the Breguet 941 (fig. 105), a four-engine, deflected-propeller slipstream transport, to determine the aircraft's performance and flying qualities (ref. 109), and to explore its operation in low-visibility instrument flight conditions (ref. 110). The program with the Japanese involved tests of their Shin-Meiwa STOL seaplane. This aircraft was also a propeller-driven four-engine deflected slipstream design that was evaluated to determine its flying qualities with and without stability and control augmentation at low speed (ref. 111). Curt Holzhauser and Hervey Quigley were the research leaders for these programs, and Bob Innis served as project pilot. Innis received the Octave Chanute Award in 1964 for his contributions to the evaluation of the performance and handling of this variety of aircraft. The team of Innis, Holzhauser, and Quigley was recognized by Ames Research Center for pioneering research in STOL aircraft performance and flying qualities with the H. Julian Allen Award for an outstanding research paper in 1972. That paper (ref. 112) still stands as the authoritative document on STOL flying qualities and operating characteristics.

Figure 105
Breguet 941 (French military transport).

Several preliminary STOL designs were investigated by NASA analytically and in wind tunnels in the late 1960s, and a few of these were eventually developed for flight evaluation under the auspices of a research aircraft project office, established within the Aeronautics Directorate at Ames in the late 1960s. During his stint at headquarters as head of NASA aeronautics, Bill Harper had been successful in convincing NASA upper management of the desirability of building proof-of-concept aircraft to validate ground-based technology in a full-scale vehicle in the flight environment. His idea was that these aircraft would provide a focus for the integration of the technologies required in the development of operational vehicles of the type and then could serve the NASA research community as a flight facility for more broad ranging research for that class of aircraft.[32] From the outset at Ames, the Center director, H. Julian Allen, and the director of Aeronautics, Russell (Russ) Robinson, provided

---

[31] Brad Wick 1998: personal communication.
[32] Bill Harper 1998: personal communication.

Figure 106
North American OV-10A Bronco rotating-cylinder flap research aircraft.

Figure 107
Boeing/deHavilland Augmentor Wing Jet STOL Research Aircraft.

strong leadership and advocacy for this idea. Harvey Allen's successor as director, Hans Mark, along with Leonard Roberts, who followed Russ Robinson as director of Aeronautics, and Brad Wick, chief of Flight Systems and Simulation, later were stalwarts in program advocacy and in securing approval and sustained support for several projects.[33] Woody Cook led the advanced aircraft project office from the start. Reference 113 by Dave Few provides insight into the workings of the project office, including a description of each aircraft program and its significance.

The OV-10A Bronco (fig. 106), which was the first of the several flight projects flown at Ames, served as the test bed for the rotating-cylinder-flap concept. The wing of the aircraft was modified to incorporate a two-segment flap located aft of the hydraulically driven rotating cylinder. The Bronco is a twin-propeller aircraft powered by two T-53 turbine engines, interconnected through a cross-shaft. The rotating cylinder in this case energized the boundary layer, thus keeping the airflow from separating from the wing flaps. At the same time, the cylinder also deflected the propeller thrust to provide a powered-lift component to the wing lift. The aircraft was tested in both the 40- by 80-foot wind tunnel and in flight, starting in mid-1971. Flight testing showed the anticipated improvement in low-speed performance, but also revealed adverse stability and control characteristics that prohibited the aircraft from being flown routinely to its full performance potential (ref. 114). In this case, the airplane became increasingly unstable longitudinally as speed decreased, a result of increased downwash on the horizontal stabilizer at the higher flap angles. Increasing the deflection of the trailing segment on the inboard flap improved longitudinal stability sufficiently to achieve a reduction in approach speed to 57 knots. James (Jim) Weiberg led the program and Bob Innis served as the project pilot.

A modified deHavilland C-8A Buffalo, the Augmentor Wing Jet STOL Research Aircraft (AWJSRA), was the research aircraft used in evaluating the augmentor wing concept and was the world's first jet STOL transport demonstrator (fig. 107). The aircraft was the first major aircraft development program of the new project office and was the focus of a joint NASA/Canadian Department of Industry, Trade and Commerce project to demonstrate the concept in the low-speed regime and in terminal-area operations. Woody Cook, Curt Holzhauser, and Hervey Quigley led the program definition and advocacy effort.[34] Once under way, the project was headed by Dave Few, with technical direction by Quigley. Bob Innis was the project pilot and was joined by Seth Grossmith, a pilot from the Canadian Ministry of Transport. The wing of the C-8 was replaced with one of reduced span that incorporated augmentor flaps, spoilers, blown ailerons, and fixed leading-edge slats. In this design, ejecting fan air between the upper and lower segments of the augmentor flaps enhanced lift. The fan flow was cross-ducted from each engine to the augmentor flaps so that the system would provide balanced lift with one engine inoperative. The engines' jet exhaust could be vectored from 6 degrees to 104 degrees below the horizontal. Extensive tests were carried out in the 40- by 80-foot wind tunnel and on the Flight Simulator for Advanced Aircraft to develop the augmentor flap and the

---

[33] Woody Cook 1998: personal communication.
[34] Woody Cook 1998: personal communication.

control system design. The flight program began in mid-1972. Nominal approach speeds of 60 knots were routine, and speeds as low as 50 knots were demonstrated. Takeoff and landing distances of less than 1000 feet over a 50-foot-high obstacle were easily achieved, as were ground rolls of 350 feet. Quigley, Dick Vomaske, and Jack Stephenson documented the aerodynamics of the augmentor flap along with the airplane's performance and stability and control characteristics in references 115–117.

Figure 108
deHavilland DHC-6 Twin Otter.

After flight tests proved the powered-lift design, a digital guidance, control, and display system (STOLAND) was installed in the Augmentor Wing to provide a capability for advanced control, as well as guidance and navigation research for STOL operations. This system provided computer control of pitch, roll, and yaw, as well as thrust and thrust deflection along with electronic head down displays for precision guidance. All had been developed in the course of several experiments in the Flight Simulator for Advanced Aircraft and in precursor flight experiments with the Convair 340 and a DHC-6 Twin Otter aircraft (fig. 108), the latter lent to Ames for this purpose by the FAA.[35] The modified Augmentor Wing was then used to evaluate flying-qualities criteria, augmented controls, and flight director concepts under the leadership of Jack Franklin and Bill Hindson, the latter representing the Canadian National Aeronautical Establishment.

Flying qualities design criteria for flightpath and speed control during landing approach and in the flare to touchdown resulted from these tests (ref. 118). The criteria were used in the Air Force's development of specifications for STOL transports and were ultimately applied to the C-17 military transport design. The data were also used by the FAA in defining airworthiness criteria for this category of civil transports. Experience with attitude, flightpath, and speed-control augmentation designs, flight director guidance, and operating procedures are documented in reference 119.

An automatic approach and landing system for STOL terminal-area operations was also developed and flown extensively. Donald Smith and DeLamar (Del) Watson headed up this effort (ref. 120) and demonstrated, for the first time, fully automatic flight for a powered-lift STOL aircraft. Gordon Hardy served as project pilot for this phase of the program with support from Bill Hindson. Later on, Luigi Cicolani and George Meyer used the aircraft for the flight demonstration of Meyer's nonlinear inverse control concept to a full flight envelope autopilot for a powered-lift aircraft (ref. 121). The initial effort with nonlinear inverse control had been carried out on the DHC-6 to prove the concept would work in flight. These efforts were the precursor of research by Meyer and several of his colleagues in which this control scheme was applied to a variety of powered-lift vehicles. The nonlinear inverse concept eventually joined classical control design methods as a contending approach for control augmentation design in the U.S. industry.

[35] Del Watson 1998: personal communication.

Figure 109
Crows Landing Naval Auxiliary Landing Field and flight research facility, Crows Landing, Calif.

Figure 110
Boeing Quiet Short Haul Research Aircraft (QSRA).

It is appropriate to note that an important contribution to the success of these flight research programs came from the development of the Crows Landing remote test facility (fig. 109). Ames had used this airfield, long a U.S. Navy auxiliary landing field, for several tests over the years, most notably for the carrier approach speed and steep descent experiments cited earlier. However, the addition of a precision tracking radar and laser system, an experimental microwave landing system, and a digital data acquisition and processing system proved to be crucial to the successful accomplishment of these STOL research projects, as well as those for V/STOL and rotorcraft and occasional other civil and military experiments. The leadership of Henry Lessing, along with key contributions by Michio Aoyagi, Michael Bondi, and the NASA and contractor team of the Avionics Research Branch, in the development and operation of this facility are to be commended.

The Quiet Short-Haul Research Aircraft (QSRA) was the last of the STOL transport designs to be carried to flight evaluations by the Ames project office (fig. 110). A major objective of this development was to achieve STOL performance at the lowest noise levels possible. Wally Deckert, Curt Holzhauser, David Hickey, and Anthony Cook were instrumental in defining the program and in having it approved.[36] This aircraft used upper surface blowing (USB) and attained short-field takeoff and landing performance that ultimately exceeded that of all the competing designs. Modified by Boeing from a deHavilland C-8A Buffalo aircraft, the QSRA featured four jet engines whose exhaust was directed over the upper surface of the wing and curved flaps. Through the Coanda effect, a portion of the propulsive force was deflected into propulsive lift while lift was further augmented by increased circulation associated with the high-velocity exhaust air flowing over the wing. Once again, this design was thoroughly developed during tests in the 40- by 80-foot wind tunnel and the Flight Simulator for Advanced Aircraft.

The first flight took place in mid-1978. John Cochrane led the project and, along with his team (shown in fig. 111 next to the original C-8A), completed the proof-of-concept phase ahead of schedule and under budget. Jim Martin was the project pilot. Dennis Riddle assumed responsibility for the first phase of the flight research program during which he, Victor Stevens, and Michael Shovlin served as principal investigators. During the initial performance and stability and control test phase, the aircraft achieved stable flight at lift levels three times those generated on conventional aircraft, although the levels of lift obtained were somewhat less than those achieved in the wind tunnel tests (ref. 122).

Noise levels of 90 EPNdB (equivalent perceived noise) at a sideline of 500 feet were obtained, the lowest achieved for any jet STOL transport design. The aircraft's noise footprint was substantially smaller than that of a comparable conventional jet transport. The QSRA further demonstrated its STOL performance by operating aboard the aircraft carrier U.S.S. Kitty Hawk without a need for catapult launch or landing arresting gear (ref. 123). A series of reduced-thrust takeoffs was performed to

[36] Woody Cook 1998: personal communication.

Figure 111

QSRA project team. From left to right: John Cochrane, Robert Price, Howard Turner, Mike Shovlin, Dennis Riddle, Al Boissevain, Dennis Brown, Patty Beck, John Weyers, Bob McCracken, Peter Patterakis, Jack Ratcliff, Al Kass, Bob Innis, Tom Twiggs (Boeing).

demonstrate the applicability of this high-lift USB technology to future powered-lift transport aircraft with more conventional thrust-to-weight ratios. Results of those tests showed substantial increases in payload or reductions in field length at the lower thrust-to-weight ratio compared to current wing/flap designs (ref. 124). Stability and control characteristics extracted in flight were analyzed and documented by Jack Stephenson in references 125 and 126. In the interest of technology transfer to potential users, an extensive flight demonstration program was carried out to introduce U.S. military and civilian pilots and operators to high-performance, quiet STOL operations with the aircraft. Later, as part of a technical exchange program with the Japanese National Aerospace Laboratory, Japanese pilots had the opportunity to fly the QSRA, and Jim Martin and Gordon Hardy in turn flew the new Japanese STOL transport, the ASKA, a four-engine, upper-surface-blown flap design.

The QSRA was later equipped with a digital, fly-by-wire control system and head-up and head-down electronic displays that used the flightpath-centered pursuit-tracking idea pioneered earlier by Dick Bray. George Meyer's nonlinear inverse control method was employed in the design of the powered-lift controls. Del Watson led the system development effort. Then the aircraft was flown in a program to evaluate integrated flightpath/airspeed controls and displays for making precision instrument approaches and landings; Gordon Hardy was the project pilot. Jack Franklin, Charlie Hynes, and Del Watson led the research team (fig. 112) through several experiments. Flying instrument approaches, the pilots were able to achieve precise control to the desired approach path and assessed the flightpath and speed command controls and HUD to produce fully satisfactory flying qualities (refs. 127 and 128). Further experiments included instrument approaches to touchdown, flown using the HUD with modifications to the control system to compensate for ground effect. These tests produced touchdown accuracies comparable to those for non-flared carrier landings (ref. 129). Influences of control augmentation and wind conditions on the precision landing capability were documented in reference 130. The Air Force developed flying qualities specifications for the C-17 transport based on the results of the earlier Ames Augmentor Wing and QSRA research flights. Pilots on the C-17 joint test team from the Air Force and McDonnell Douglas flew the QSRA in a program run by Hynes and Hardy to evaluate flightpath control augmentation and the head-up display. Results of these tests were eventually incorporated into the C-17 design. Hynes and Hardy also used their experience with display design for STOL aircraft to assist Lockheed with a display application to the MC-130E.

Figure 112
QSRA research team. Front row: Jim Ahlman, Bob Innis, Del Watson, Jim Lesko, Lee Mountz, Mike Herschel, Tom Kaisersatt, Jack Stephenson. Back row: Dennis Riddle, Nels Watz, Jack Franklin, Gordon Hardy, Bob Hinds, Charlie Hynes, Richard Young, Jim Martin, Joe Eppel, John White, Bob America, Hien Tran, Bill Bjorkman.

The last research flights with the QSRA were jump-strut tests, conducted by Joseph Eppel and flown by Martin and Hardy. In these flights, the nose landing gear hydraulic system was used to initiate the nose-up rotation of the aircraft during takeoff roll, permitting a further reduction of takeoff distance. In addition to research, the QSRA demonstrated the capability of USB powered lift at the 1983 Paris International Air Show and the 1986 Expo '86/Abbotsford Air Show. The flight from Moffett Field to Paris Le Bourget and return, carried out at about 200 knots over several legs along a nearly great circle route, was an accomplishment in itself. A chronology of the QSRA program and its accomplishments has been documented by Cochrane and Riddle and their team in references 131 and 132.

As a footnote to the guidance and control research carried out on these aircraft, Sperry Flight Systems made particular use of the experience gained from flying these digital systems in state-of-the-art flight computers. Evolutions of those computers found their way on to the space shuttle trainer aircraft, the AV-8B, the DC-10 refueling system, and the MD-80.[37]

Figure 113
Six-Degree-of-Freedom Simulator.

V/STOL RESEARCH

As was mentioned in the Introduction, Ames became the center for V/STOL research when the NACA was absorbed into NASA. This decision was a consequence of Ames' experience with low-speed aircraft flying qualities and of the availability of the 40- by 80-foot wind tunnel for low-speed, full-scale aircraft testing.[38] The technical issues associated with V/STOL concerned the means by which an aircraft could be configured to achieve acceptable hover and cruise flight performance and be able to transition between the two flight regimes with ease. Controllability was always a concern at low speed, because conventional aerodynamic surfaces were no longer effective and the propulsion system then became the only source of control. Many different aircraft arrangements were explored in the wind tunnel and a few eventually made their way into flight. The V/STOL program also led to the development of two new flight simulators that were used extensively in conjunction with Ames V/STOL flight research. From his position as division chief, Bill Harper encouraged the development of the Six-Degree-of-Freedom hover flight simulator (fig. 113), the first large motion device at Ames. Built in the mid-1960s, it was an open cockpit arrangement that operated within an 18-foot cube. In the mid-1970s, the design of a much larger motion platform, the Vertical Motion Simulator (fig. 114), was initiated to support the Center's powered-lift and rotorcraft programs. This facility, which is still the world's largest motion simulator, combined 60 feet of vertical travel and 40 feet of either longitudinal or lateral travel with high-response rotational freedom to produce cockpit motion with the closest resemblance to flight of any ground-based facility. It has been used in the development of a number of aircraft configurations and flight control systems, and has been tightly linked to all of Ames' recent flight research programs.

Figure 114
Vertical Motion Simulator.

[37] Del Watson 1998: personal communication.
[38] Bill Harper 1998: personal communication.

Figure 115
Ryan VZ-3RY Vertiplane.

Figure 116
Bell X-14A (VTOL experimental aircraft).

Ames' experience with V/STOL configurations in the late 1950s came from flying the VZ-3, X-14, and XV-3, along with the VZ-2 and VZ-4, and formed the basis for early attempts to define flying qualities criteria and to gain an understanding of operational techniques for these aircraft (refs. 133 and 134). The VZ-3RY (fig. 115) used deflected propeller slipstream to augment wing lift, and a form of engine exhaust gas reaction control for low-speed pitch and yaw control. Ames added full-span slats to the wing to increase its lift. Tests were carried out in the 40- by 80-foot wind tunnel to define its performance, stability and control, and handling characteristics. With any wind, the aircraft could nearly hover out of ground effect, but it was ungainly and difficult to control in the presence of gusts. Its flying qualities and control characteristics were explored and are documented in reference 135. Howard Turner led the project and Glen Stinnett and Fred Drinkwater did most of the flying. The aircraft was lost when Stinnett ran out of nose-down control at low power and the aircraft pitched inverted and crashed into San Francisco Bay. He was able to eject and survived to continue his career in Ames flight test. The aircraft was subsequently rebuilt to complete the test program.

The first jet VTOL aircraft to be flown at Ames was the X-14, a configuration developed by Bell Aerospace for the Air Force from a Beech T-34 wing and tail. It was initially powered by two Bristol-Siddeley Viper engines exhausting through cascade thrust deflectors. Ames developed an analog variable stability system for the aircraft for use in conducting flying qualities investigations in hover. Frank Pauli was in charge of the variable stability system design, which is described in reference 136. The aircraft was also fitted with General Electric J-85 turbojet engines at that time to increase thrust margins for hover. It was then redesignated as the X-14A (fig. 116). Extensive flight testing was led by Stew Rolls to investigate a range of flying qualities in hover; those flight tests resulted in criteria for longitudinal, lateral, and directional control power, sensitivity and damping (refs. 137 and 138). Height-control requirements were developed from X-14 flight data and from simulator results by Ron Gerdes (ref. 139). Tests of lateral thrust vectoring control, as a means of achieving lateral translation without banking the aircraft, were carried out by Terrell Feistel and Emmett Fry (ref. 140). Drinkwater and Gerdes were the principal pilots throughout this program. Research with this aircraft, in conjunction with a number of related experiments on the new Six-Degree-of-Freedom hover simulator, contributed to the military flying qualities specification for V/STOL aircraft and played an important role in the control system development of the Hawker P.1127, a British V/STOL tactical fighter that was developed into the Harrier, the western world's only operational fixed-wing V/STOL strike fighter. Additionally, in 1965 the X-14 was flown by Neil Armstrong to evaluate control characteristics in vertical flight that would be representative of the Apollo lunar lander during final descent to landing on the Moon. Drinkwater's contributions in flight testing these V/STOL aircraft were recognized when he received the Octave Chanute Award in 1964.

The aircraft was again modified, under the direction of Richard (Dick) Greif and Terry Gossett, to install a digital variable stability system and uprated GE J-85 engines; it then became known as the X-14B (fig. 117). Ron Gerdes then assumed sole responsibility as the project pilot. Flight and simulation experiments during that period were conducted by Lloyd Corliss and Dick Greif to establish criteria for pitch and roll attitude command concepts, which had become the control augmentation of choice for precision hover (ref. 141). During a later experiment phase in 1981, the aircraft made a hard landing as a consequence of a lateral control software design flaw that led to a pilot-induced oscillation. It was never flown again. The test team, assembled for the aircraft's twentieth anniversary in 1977, appears in figure 118.

As in the case of STOL research, Ames V/STOL expertise led to interaction with aeronautical establishments in other countries that were pursuing this technology. Contacts overseas came through the North Atlantic Treaty Organization's Advisory Group for Aeronautical Research and Development (AGARD), often as a

Figure 117
Bell X-14B (VTOL experimental aircraft).

Figure 118
X-14 team. Front row: Fred Drinkwater, Jim Meeks, Lonnie Phillips, Jim Kozalski, Vic Bravo. Second row: Bill Carpenter, Sid Selan, Dick Gallant, Terry Stoeffler. Third row: Ron Gerdes, Lloyd Corliss. Fourth row: Cy Sewell, Dick Greif, Ed Vernon, Lee Jones. Fifth row: Dan Dugan, Jim Rogers, Dave Walton, Terry Feistel. Back row: Frank Pauli, Seth Anderson. Not pictured: Terry Gossett, Bob Innis, Stew Rolls, Lawson Williamson.

Figure 119
Hawker P.1127 (U.K. experimental
V/STOL fighter).

Figure 120
Dornier DO-31 (German experimental
VTOL transport).

Figure 121
Bell XV-3 (experimental tilt rotor).

consequence of Seth Anderson's membership on the V/STOL panel. Thus, Ames became involved in flight research and evaluations with two noteworthy jet-lift developments in England and Germany, the P.1127 and the Dornier DO-31. Before the first flight of the P.1127 in England, the Hawker test pilots, Bill Bedford and Hugh Merewether, came to Ames to fly the X-14 in order to acquaint themselves with handling a jet V/STOL aircraft in hover. The two went away with an appreciation of the skill required to hover an unstabilized vehicle and of the control sensitivities necessary to do so. After the P.1127 program was under way, Fred Drinkwater had an opportunity to fly that aircraft to explore its flying qualities in transition and hover (fig. 119). The DO-31 program was established by Woody Cook, Paul Yaggy from the Army organization at Ames, and Jack Brewer from NASA Headquarters. It provided for simulation evaluations of the aircraft at Ames, and flight tests by a NASA Ames and Langley team at the Dornier facility outside of Munich. Curt Holzhauser and Bob Innis were assigned as the Ames representatives to evaluate this jet VTOL transport.[39] The aircraft had a mixed-propulsion arrangement, including eight wing-tip mounted lift engines, and two wing-pod-mounted lift-cruise engines whose thrust could be deflected for transition and hover (fig. 120). The flight program was aimed at acquiring experience with this class of V/STOL aircraft to prepare for their then-anticipated development for commercial service. Flight tests generated data on the performance, flying qualities, and operating characteristics of the DO-31 in transition and during approach and vertical landing, including simulated instrument flight. The aircraft exhibited a broad transition performance envelope and good attitude-control characteristics with the attitude-command augmentation system. A primary deficiency concerned the multiplicity of controls the pilot was required to manipulate for flightpath and airspeed control during the deceleration to hover, particularly under instrument meteorological conditions (ref. 142).

One concept among many envisioned by the Army in its efforts to combine the hover performance of the helicopter with the cruise flight capability of propeller-driven aircraft was the tilt rotor. The tilt-rotor configuration uses large-diameter rotors mounted on wing-tip nacelles to hover with a significant payload. For cruise flight, the rotors tilt forward so that they operate as propellers to generate the thrust necessary for high speeds. The first successful tilt-rotor aircraft, the XV-3 (fig. 121), was produced by Bell Helicopter for the Air Force and Army and went through extensive development testing in the 40- by 80-foot wind tunnel before being flown by the Air Force at Edwards Air Force Base and subsequently at Ames. The Air Force tests were led by Wally Deckert, who, prior to joining Ames, was a member of the Air Force team.[40] These tests, along with an earlier series at Bell, identified a rotor/nacelle/wing whirl mode instability that limited the flight envelope to 130 knots, severely restricting the aircraft's desired performance envelope. When the aircraft arrived at Ames, Hervey Quigley carried out the research and Don Heinle and Fred Drinkwater conducted much of the test flying. In the Ames tests, flapping of the teetering rotors during maneuvers introduced moments that reduced damping of the longitudinal

[39] Woody Cook 1998: personal communication.
[40] Woody Cook 1998: personal communication.

and lateral-directional oscillations to near zero at speeds approaching 140 knots (ref. 143). Despite these problems and despite being under-powered and limited in payload, the XV-3 proved the capability of the tilt rotor to perform in-flight conversions between the helicopter and the airplane modes. Analytical studies and test data showed that the XV-3 design had a substantial transition flight envelope. However, the rotor dynamics and flight control issues needed to be resolved for the promising attributes of the tilt-rotor concept to be achieved. These problems were attacked by Bell through extensive analytical studies and scale model experiments leading to another round of 40- by 80-foot wind tunnel tests. Results produced an understanding of the physics of rotor-pylon dynamics and validated methods for assuring stability.[41]

Figure 122
Bell XV-15 Tilt Rotor Research Aircraft.

The demonstration of fundamental tilt-rotor capabilities with the XV-3 flight tests resulted in the formation of a NASA/Army joint project at Ames to further develop tilt-rotor technology through contracted and in-house analytical and experimental efforts. This work culminated in the most significant demonstrator aircraft program that Ames has pursued, the design and construction of two XV-15 Tilt Rotor Research Aircraft (fig. 122). The XV-15 was the first proof-of-concept aircraft built as an entirely new airframe to Ames' specifications. Leadership during the program definition came from Wally Deckert, who had moved to Ames by then, along with Mark Kelly and Demo Giulianetti, and from Paul Yaggy, Dean Borgman, and Kipling (Kip) Edenborough of the Army Laboratory at Ames.[42] Again, Bell Helicopter produced the flight vehicle. Once the program was under way, Dave Few served as the project manager followed by Army LTC James Brown and later by John Magee.

The aeronautical facilities at Ames played an important part in the design and test of these aircraft, including wheel-pod drag tests in the 7- by 10-foot wind tunnel, rotor performance and dynamics tests in the 40- by 80-foot wind tunnel, and a number of control systems development piloted simulations in the Flight Simulator for Advanced Aircraft. Prior to flight envelope expansion, the first XV-15 (NASA 702) was tested in the 40- by 80-foot wind tunnel in mid-1978 for a preliminary evaluation of the aircraft's aerodynamic and aeroelastic characteristics.

At the completion of envelope expansion flight tests at the Dryden Flight Research Center by the Ames project team, the second aircraft (NASA 703) was delivered to Ames in mid-1981. It became the subject of a series of technology development activities over the next two decades. Laurel Schroers led the flight research program, and he, Gary Churchill, Marty Maisel, and Jim Weiberg served as principal investigators. Daniel (Dan) Dugan, Ron Gerdes, George Tucker, LTC Grady Wilson, and LTC Rick Simmons were program pilots at various stages. The project team is pictured in figure 123. The flight activity included flying qualities and stability and control evaluations, control law development, side-stick controller tests, performance evaluations in all flight modes, acoustics tests, flow surveys, and documentation of its loads, structural dynamics, and aeroelastic stability characteristics (refs. 144–146).

[41] Kip Edenborough 1998: personal communication.

[42] Woody Cook 1998 personal communication; Bill Snyder and Marty Maisel 1998: personal communication.

Speeds in cruise flight exceeding 300 knots were achieved and these tests showed that the wing-pylon whirl-mode instability had been eliminated within the flight envelope (ref. 147). Comparisons of wind tunnel and flight results are presented in reference 148. A large digital database from the program was maintained on Ames' computer facilities and made available on-line for use by U.S. industry and the military services. Advanced Technology Blades, developed by Boeing under the leadership of Marty Maisel, were flown on the aircraft to advance the technology for rotor-blade design.

The aircraft's large transition envelope and good flying qualities were found to make it easy for pilots to operate in any flight regime from cruise to hover. Operational demonstrations were performed for over 100 military and industry pilots, and included nap-of-the-Earth flight, air-to-air combat, aerial refueling, and launch and recovery aboard an aircraft carrier. The aircraft also performed flight displays and was

Figure 123

XV-15 project team. Front row: Mike Bondi, Dan Dugan, Shorty Schroers, Wally Deckert, Marty Maisel, Violet Lamica, Robby Robinson, Demo Giulianetti. Back row: Jerry Bree, Gary Churchill, Dave Few, Jerry Barrack, Kip Edenborough, Jim Lane, Mike Carness, Dave Chappel, Duane Allen. Not pictured: Woody Cook, Jim Weiberg, Dean Borgman, Jim Brown, John Hemiup, Al Gahler, Ron Gerdes, Cliff McKiethan, Bill Snyder, Rick Simmons.

exhibited at the Paris Air Show in 1981. In recognition of the significant contributions of the XV-15 program, the project team received the American Helicopter Society's Grover E. Bell award in 1980. The validation of tilt-rotor analytical methods resulting from the XV-15 flight program provided sufficient confidence in the technology for the initiation of the JVX program, which led to the Bell-Boeing V-22 Osprey program by the U.S. Marines. The Ames flight program was terminated following the Advanced Technology Blade Project when, during subsequent acoustic tests, the aircraft was subjected to severe vibratory loads in conjunction with a blade-root cuff failure. NASA 703 was bailed to Bell to continue tilt-rotor technology development and demonstration flying. In addition, it provided direct support to the NASA Short Haul Civil Tilt Rotor Project at Ames. The Bell model 609 civil tilt rotor, designed to carry six to nine passengers, directly evolved from the 15,000-pound XV-15.

Lift fans were another powered-lift generating device that received thorough scrutiny by V/STOL configuration developers and by Ames aerodynamicists in the 40- by 80-foot wind tunnel. Tests in and out of ground effect of various wing and inlet configurations, exit-vane designs, nose fans, and control devices were carried out. David Hickey led these investigations at Ames. A flight vehicle came out of this work in the form of the Army-sponsored Ryan XV-5A, which used two J-85 engines either for cruise thrust or, with its exhaust flow diverted, to drive tip turbines on two wing-mounted lift fans and a nose-mounted pitch fan.[43] Movable vanes in the exit plane of the wing fans could either deflect or spoil fan thrust. Army flight tests in the mid-1960s were sufficiently encouraging, despite a marginal transition corridor and lack of short takeoff capability, that the XV-5A was rebuilt following a fatal crash.[44] Research at Ames began after the reconstruction of the damaged airframe into the XV-5B (fig. 124). The program included aerodynamic, acoustics, and flying qualities evaluations of the lift-fan configuration and an investigation of the transition-to-hover using different configurations and control techniques. Correlation of the flight-measured aerodynamics and acoustics characteristics with the earlier wind tunnel test results were reported in reference 149. Flightpath control procedures were complex, and it was difficult to find a compromise control procedure for flying a precision approach (refs. 150 and 151). Charlie Hynes was the technical leader of these flight tests, which were flown extensively by Ron Gerdes.

The last of the V/STOL research aircraft flown at Ames was the YAV-8B Harrier. It was lent to Ames by the U.S. Marines in 1984 so that Ames could carry out a program of advanced controls and displays research that the Marines anticipated would be applied to the next generation of V/STOL fighter aircraft. The flight research effort was based on results of an extensive program carried out by Vern Merrick on the Vertical Motion Simulator to screen and develop promising control and display concepts. This research followed from that conducted earlier on the X-14 and in simulation experiments conducted by Lloyd Corliss on translational rate command systems. It was motivated by the desire of the Navy and Marines to operate aboard assault carriers and even destroyers in adverse weather and sea conditions.

Figure 124
Ryan XV-5B ("fan-in-wing" VTOL research aircraft).

---

[43] Woody Cook 1998: personal communication.
[44] Ron Gerdes 1998: personal communication.

Figure 125
McDonnell Douglas YAV-8B V/STOL Systems Research Aircraft (VSRA).

The aircraft, the remaining prototype for the AV-8B, incorporated the AV-8B wing, a modified engine inlet and cold exhaust nozzles, and under-fuselage lift-improvement devices in an otherwise stock AV-8A fuselage and empennage. The aircraft was powered by a single Rolls-Royce Pegasus turbofan engine. It was modified into the V/STOL Systems Research Aircraft (VSRA, fig. 125) with the installation of digital fly-by-wire controls for pitch, roll, yaw, thrust magnitude and thrust deflection, and programmable electronic head-up displays. Del Watson and John D. Foster led the team that developed this highly complex system, with the frequent consultation of Merrick. Charlie Hynes and Ernesto (Ernie) Moralez carried out the software development, K. C. Shih specified the servo requirements, and Nicholas (Nick) Rediess completed the hardware implementation. This system development and aircraft modification was performed almost entirely by Watson's team, which did the design, system integration, and installation. Alan Page served as the aircraft manager and as the contact with the Marines and Navy on all aspects of Harrier operation and maintenance. Foster provides background for the program and an overview of the system in reference 152.

An extensive flight program was then carried out on the VSRA through transition-to-hover and vertical landing to evaluate the candidate control schemes. These experiments identified the flying qualities trade-offs for the range of control augmentation concepts, and demonstrated that fully satisfactory flying qualities could be achieved with decoupled flightpath and longitudinal command controls during a continuously decelerating approach to hover. Further, a three-axis translational rate command system proved satisfactory for precision hover and vertical landing. In addition, the control authority used by each of the designs was documented for the designers' use. Jack Franklin led this phase of the research and reported the results in reference 153. Advanced guidance and navigation displays based on the flightpath-centered pursuit-tracking idea pioneered by Dick Bray were also evaluated on the aircraft. They were found to offer excellent guidance for a complex approach path and to give the pilot the ability to achieve precise hover positioning for the vertical landing (ref. 154). Daniel Dorr, Ernie Moralez, and Vern Merrick were responsible for this work. Ron Gerdes, Michael Stortz, and Gordon Hardy served as project pilots over the course of the research program. The project team is shown in figure 126. Outside pilot participation came from the U.S. Marines, the U.K. Royal Air Force, McDonnell Douglas, and Rolls-Royce.

Results of the flight tests, in combination with simulation experiments on the Vertical Motion Simulator, were used by Franklin to develop flying qualities criteria and control system and display designs for future short takeoff and vertical landing (STOVL) fighter aircraft as part of the Joint Strike Fighter program. The displays are also being installed in the AV-8B Harrier to improve precision and reduce the pilot's workload during recovery aboard ship at night. Flight tests were also conducted with the basic YAV-8B to establish reaction control bleed flow in low-speed flight to measure jet-induced ground effects, and to conduct infrared measurements of hot gas

Figure 126

VSRA team. Front row: Dave Walton, Seth Kurasaki, Bill Laurie, Jim Ahlman, Nels Watz, Del Watson, Terry Stoeffler, Linda Blyskal, Ed Hess, Manny Irrizarry, Mike Stortz, Bruce Gallmeyer. Second row: Dave Nishikawa, Stan Uyeda, Trudy Schlaich, Tom Kaisersatt, John Foster, Nick Rediess, Kent Shiffer, Paul Borchers, Mike Casey, Sterling Smith, Charlie Hynes, Vern Merrick, Jack Franklin. Back Row: Thad Frazier, Eric Weirshauser, Steve Timmons, Brian Hookland, Joe Paz, Ken Christensen, Jack Trapp, Bill Bjorkman, Ernie Moralez, Joe Konecni.

flow fields near the ground for correlation with computational fluid dynamics predictions. The bleed-flow tests provided a detailed look at the attitude-control power used by the pilot during maneuvers in all the phases of jet-borne flight. Results of some of these efforts are presented in references 155–157. Flight research was completed with this aircraft in late 1997.

Figure 127
Boeing X-36 (tailless, unmanned research vehicle).

The last of Ames' project aircraft was the X-36, an unmanned, tailless, scale model of an advanced, highly agile, fighter configuration. Even though it is not a V/STOL aircraft, it is included here to complete the picture of the efforts of the Ames aircraft project office. The objective of the X-36 program was to demonstrate that a tailless aircraft could achieve the maneuverability and agility of current-class fighters at angles of attack up to stall without the directional stabilization and control power provided by vertical tails.[45] The aircraft, at 28 percent scale, was developed for Ames by Boeing's Phantom Works. As can be seen in figure 127, it is a wing-canard configuration without vertical stabilizers. It is 18 feet long with a 10-foot wing span, weighs 1245 pounds, and is powered by a Williams Research F112 turbofan engine that produces 700 pounds of thrust and includes thrust-vectoring control. Under normal operation, the aircraft was flown remotely by a pilot sitting in a ground station using a head-up display. It could also be flown through an autopilot and was also capable of autonomous operation. Flight tests were carried out at the Dryden Flight Research Center. The aircraft was flown to angles of attack up to 40 degrees, and it demonstrated excellent stability and maneuverability up to those conditions. Rodney (Rod) Bailey was the program manager, and Mark Sumich led the project team. The NASA and Boeing team won the American Institute of Aeronautics and Astronautics Aerospace Design Engineering Award in recognition of the program's contributions.

Ames flight operations personnel as they appeared in 1984 can be seen in figure 128.

---

[45] Rod Bailey and Lloyd Corliss 1998: personal communication.

Figure 128

Flight operations personnel circa 1984. Front row: Wally Stahl, Ron Gerdes, Pat Morris, Dick Gallant, Warren Hall, Glen Stinnett, Frank Kosik.

Second row: Jack McLaughlin, Marc Betters, Fred Drinkwater, Nancy Lowe, Kathleen Burns, Vicki Rodriguez, Nancy Bouchet.

Back Row: Dan Dugan, Gordon Hardy, Bob Innis, Grady Wilson, Casey Call, Tex Ritter, Jim Martin, Mike Landis, George Tucker.

Rotorcraft Research

Rotorcraft flight research began in earnest at Ames in the early 1970s in conjunction with the newly established program between NASA and the U.S. Army in rotorcraft technology and further to support NASA's emphasis on civil rotorcraft. This work accelerated in the late 1970s with the arrival of several aircraft from NASA Langley when rotorcraft research was consolidated at Ames. The flight research activity initially concentrated on control and handling issues for terminal-area operations under adverse weather conditions and was pursued along with an extensive ground-based simulator program to develop the control systems for flight. Later on, rotor aerodynamics, acoustics, vibration, loads, advanced concepts, and human factors research would be included as important elements in the joint program activity.[46] The various rotorcraft in operation at Ames are noted in table 9.

After its arrival at Ames, the UH-1H (fig. 129) was flown extensively in a series of experiments to develop and evaluate control systems for fully automatic flight for helicopters. This work was driven by the need to develop a database for navigation and guidance concepts for instrument flight operations. A fully automatic digital flight and guidance system known as V/STOLAND that had conventional autopilot capabilities including autoland was developed for the program under the leadership of Fred Baker. The system used Kalman filtering for extracting aircraft position and inertial velocities from multiple ground-based and on-board sensors based on the earlier navigation system research.[47] Complex approach profiles consisting of helical descending flightpaths were investigated as a means of confining the operational airspace for the helicopter and segregating it from conventional transport operations at crowded airports (ref. 158). A variety of approach profiles and procedures were also examined for manual operations to provide the FAA background on the aircraft system requirements and limitations for these operations (ref. 159). Principal engineers for these experiments were George Xenakis, John D. Foster, and Harry Swenson.

Figure 129
Bell UH-1H V/STOLAND helicopter.

TABLE 9. AIRCRAFT USED FOR ROTORCRAFT RESEARCH		
Aircraft Name	Arrival or First Flight Date	Departure Date
H-23C (USA 56-2288)	November 3, 1958	April 28, 1959
UH-1B (USA 62-1908 NASA 732)	October 14, 1970	February 10, 1980
UH-1H (USA 69-15231 NASA 733)	May 4, 1974	April 20, 1988
YO-3A (USA 69-18010 NASA 718)	April 27, 1977	June 27, 1997
SH-3G (Bu. No. 149723 NASA 735)	November 9, 1977	July 27, 1993
UH-1H (USA 64-13628 NASA 734)	March 1, 1978	September 29, 1993
AH-1G (USA 66-15248 NASA 736)	March 1, 1978	May 23, 1985
RSRA (72-002 NASA 741)	February 12, 1979	October 10, 1991
CH-47B (USA 66-19138 NASA 737)	August 14, 1979	September 20, 1989
RSRA (72-001 NASA 740)	September 29, 1979	October 10, 1991
JAH-1S (USA 77-22768 NASA 730)	May 8, 1985	July 5, 1988
NAH-1S (USA 70-15979 NASA 736)	November 10, 1987	
UH-60 (USA 82-23748 NASA 748)	September 22, 1988	
JUH-60 (USA 78-23012 NASA 750)	September 23, 1989	

[46] Bill Snyder 1998: personal communication.
[47] John D. Foster 1998: personal communication.

Figure 130
Bell AH-1G Huey Cobra.

Figure 131
Sikorsky Rotor Systems Research Aircraft
(RSRA) helicopter configuration.

Figure 132
Sikorsky Rotor Systems Research Aircraft
(RSRA) compound configuration.

The aircraft was also used in a series of flights to investigate flying qualities criteria for nap-of-the-Earth operation and certification criteria for civil helicopter operations (refs. 160 and 161). Lloyd Corliss and Victor Lebacqz led these respective programs. Results identified the important features of the helicopter's response for low-altitude maneuvering and the longitudinal stability, control augmentation, and guidance and control displays necessary for civil instrument flight operations. Subsequently, the first demonstration on a helicopter of automatic control laws that used the nonlinear inverse method of George Meyer was conducted on the UH-1H. With these automatic controls the aircraft was flown from takeoff to cruise flight, then through the helical descent back to hover and landing, a first for this approach. Dan Dugan and Ron Gerdes were the project pilots for the UH-1H research.

The AH-1G White Cobra, the original NASA 736 (fig. 130), had originally been flight tested at Langley Research Center to examine the effects of different aerodynamic blade designs on rotor performance and loads. On its arrival at Ames, the Tip Aerodynamics and Acoustics Test was initiated to obtain extensive aerodynamic and load measurements to provide a better understanding of prediction methods and of the underlying physical phenomena for this rotor.[48] The highly instrumented rotor blades and instrumentation package used by the U.S. Army for the previous Operational Loads Survey tests were obtained. Additional absolute pressure instrumentation was added to the rotor tip to increase the number of radial stations measured from five to eight. This resulted in a total of 188 pressure transducer measurements on one rotor blade and an additional 126 measurements on the other blade and rotor hub. Detailed aerodynamic and performance measurements were made (refs. 162 and 163) and acoustic measurements were also obtained in flight with the YO-3A and with a ground array. Gerald Shockey was the project leader; Jeff Cross and Michael Watts were the engineers responsible for the tests, and Army LTC Robert (Bob) Merrill served as the project pilot.

Helicopter test beds for investigating new rotor concepts in flight were developed under a NASA/Army program at Langley Research Center and later transferred to Ames to be used as flight research facilities. Two vehicles, built by Sikorsky Aircraft, were known as the Rotor Systems Research Aircraft (RSRA), one in a helicopter configuration (fig. 131), the other a compound helicopter (fig. 132). They were designed to be fully capable of flight in three different modes: helicopter, compound (with or without wings), and fixed wing with no rotor.[49] These aircraft were powered by two T-58-GE-5 turboshaft engines through an S-61 main transmission. The compound configuration added two TF-34-GE-400A turbofans as auxiliary engines and a large wing including fixed-wing control surfaces. During the first several years at Ames, the aircraft flight envelopes were expanded to the design limits. These aircraft were intended to test new rotor concepts at full scale in the flight environment at conditions that could not be achieved in a wind tunnel. The compound aircraft was equipped with 14 load cells to measure main-rotor thrust, torque, and drag, wing lift and drag, tail-rotor thrust, and auxiliary engine thrust; the helicopter

[48] William (Bill) Bousman 1998: personal communication.
[49] Bill Snyder and Gregory Condon 1998: personal communication.

version was instrumented to measure main-rotor thrust, torque, drag, and tail-rotor thrust. Both aircraft were equipped to measure over 500 other aircraft and rotor state, structural, and acceleration parameters. Representative results from the flight-test program with the two aircraft are noted in references 164–167. The helicopter version was diverted as the base airframe for the X-Wing concept development.

The compound configuration continued in flight research and provided critical support to the X-Wing project. One contribution came in a flight test by the Ames team at Dryden, where the aircraft was flown without the rotor system. This flight mode was required for the aircraft to test high-risk rotor configurations, which might have to be separated from the aircraft as a result of system failure or instability. The pyrotechnic blade-separation system was used for both the RSRA and X-Wing. Gregory Condon led the program at the outset, followed at a later date by William (Bill) Snyder. John Burks, Ruben Erickson, and Ed Seto served as test directors for the two aircraft. Warren Hall was the project pilot throughout the program; LTC Bob Merrill also participated as a program pilot during the early stages. The RSRA compound configuration was placed in flyable storage in 1986 after an extensive internal assessment determined that the most cost-effective way to meet industry's needs for modern rotor air-load data was through tests with a UH-60 airframe.

Boeing's CH-47B Chinook was further developed by the Army and NASA as a variable stability helicopter at the NASA Langley Research Center. It was transferred to Ames in 1979 in support of Ames' newly assigned lead role for NASA rotorcraft research. Full-authority electrohydraulic actuators were driven originally by signals from a large on-board TR-48 analog computer. At Ames, the aircraft (fig. 133) was modified to include two digital flight computers, a programmable force-feel system, and a programmable color cathode-ray tube display. This in-flight simulation capability was the catalyst for a variety of flight experiments that ranged from investigations in support of a new flying qualities specification for military rotorcraft that was developed primarily at Ames, to the evaluation of advanced multi-input, multi-output control laws. In all cases, exploratory development of the criteria and advanced control laws was performed on the Vertical Motion Simulator before moving into flight. Experiments to further explore control during flight near the terrain showed the significance of control and response cross-coupling on flying qualities (ref. 168). Height-control requirements during bob-up maneuvers were established in the experiment reported in reference 169.

As a basis for developing advanced control laws, fundamental work was carried out on the modeling of the rotorcraft, to include the effects of high-order influences of the rotor system and to determine the sophistication of models required for control design (ref. 170). Representative of the advanced control activities were the efforts carried out on model-following systems and the effect of high-order system dynamics on the ability to tightly control the aircraft's dynamic response (refs. 171 and 172). Some of the advanced controls research was also performed in close association with

Figure 133
Boeing CH-47 Chinook in-flight simulator.

Stanford University (ref. 173). Advanced displays were evaluated to determine the information required and the symbol dynamics necessary to precisely hover and land the aircraft (refs. 174 and 175).

A number of individuals led the various experiments, including Robert Chen, Bill Hindson, Kathryn Hilbert, Douglas Watson, Michelle Eshow and Jeffery Schroeder. Hindson and George Tucker performed the bulk of the evaluation flying. Figure 134 shows the aircraft's research team. The aircraft was returned to the Army in 1989 for remanufacture to a CH-47D model. As a consequence of the expertise Ames had acquired from these wide-ranging flight and simulation activities, a team led by Mark Tischler took part in an extensive investigation of the flying qualities of Boeing's Advanced Digital Optical Control System demonstrator, a highly modified UH-60A helicopter. Their work pointed out areas of agreement between flight results and the original design guidelines for these advanced systems, as well as those aspects needing improvement (ref. 176).

The in-flight and ground-based simulation research on flying qualities for nap-of-the-Earth maneuvering conducted by the Army and NASA team led to a collaborative activity with counterparts in the German aeronautical research establishment, the

Figure 134
Boeing CH-47 research team. From left to right: Greg Condon, Emmett Fry, George Tucker, Gale Kaplan, Katie Hilbert, John Wilson, Grady Wilson, Dave Nishikawa, Bill Hindson, Al Parker, Bob Chen, Vic Lebacqz.

DFVLR. As part of the Army's memorandum of understanding with the Germans, Ames engineers and pilots took part in flight programs carried out by the Germans at their military flight test center at Manching in Bavaria.[50] From 1981 through 1984, Ron Gerdes went to Germany to fly their UH–1D and BO–105C helicopters over slalom courses set up to enable assessment of flying qualities in aggressive maneuvers near the terrain. Edwin Aiken was the Army's principal investigator for those tests. Later, LTC Grady Wilson and LTC Rick Simmons, along with Chris Blanken, now the program's technical leader, took part in flights on the BO–105C and, starting in 1989, with the German's new variable stability BO–105, to pursue investigations in more detail. This phase of the program continued until 1993.

A UH–60A Black Hawk (NASA 748, fig. 135) with conventional structural instrumentation installed on the blades was tested in 1987 at Edwards AFB under Ames' sponsorship as part of the Modern Rotor Aerodynamic Limits Survey. This effort was led by Jeff Cross; a summary is reported in reference 177. Sikorsky Aircraft was contracted to build a set of highly instrumented blades for the Black Hawk test aircraft: a pressure blade with 242 absolute pressure transducers and a strain-gauge blade with an extensive suite of strain gauges and accelerometers. This aircraft was transferred to Ames in 1988 and integration of a data system for the highly instrumented blades was started. The required high-bandwidth data of the blade pressure measurements resulted in a 7.5 megabit data stream from the rotor, a capability that was beyond the state of the art at that time. A number of approaches to obtaining the high data rate were attempted and success was finally achieved. Approximately 30 gigabytes of data were obtained in 1993–94 and installed in an electronic database that was immediately accessible to the domestic rotorcraft industry.[51] A number of results are documented in references 178–180. Ed Seto was the project manager at the outset, and William (Bill) Bousman and Robert Kufeld led a large research team that carried out this complex project (fig. 136). The project pilots were Rick Simmons and Munro Dearing. In 1995, the air-loads project team was honored with the Grover E. Bell Award by the American Helicopter Society in recognition of its contribution to the understanding of this complex technical area.

This Black Hawk was also used in a test program to develop and demonstrate a method for identifying system stability and flying qualities for slung-load operations.[52] The slung load consisted of an instrumented 8- by 6- by 6-foot cargo container. Tests focused on the characteristics of longitudinal and lateral axes with results computed as the flight tests progressed. Mark Tischler's system-identification software was used to compute flying qualities parameters, control system stability margins, and characteristic roots for the load pendulum (ref. 181). Results were also used to validate a slung-load simulation model. Tischler and George Tucker developed the program as part of a U.S./Israeli collaborative effort on rotorcraft aeromechanics and man-machine integration. Luigi Cicolani conducted the research with Rick Simmons and LTC Chris Sullivan as the project pilots.

Figure 135
Sikorsky UH-60 Air-Loads Research Aircraft.

---

[50] Ron Gerdes and Chris Blanken 1998: personal communication.

[51] Bill Bousman 1998: personal communication.

[52] Luigi Cicolani 1998: personal communication.

Figure 136

Sikorsky UH-60 air-loads research team. Front row: Frank Pichay, Jim Phillips, Karen Studebaker, Stan Uyeda, Munro Dearing, Rick Simmons, Mario Garcia, Anna Almaraz, Allen Au, Frank Pressbury, Bob Kufeld, Marianne Kidder, Nancy Bashford, Jack Brilla, Dwight Balough, Chico Rijfkogel, Paul Aristo. Back row: Tom English, Dick Denman, Patrick Brunn, Tom Reynolds, Bud Billings, Paul Espinosa, Bill Bjorkman, Chee Tung, Leonard Hee, Bill Bousman, Tom Maier, Ron Fong, Steve Timmons, Jeff Cross, Colin Coleman, Paul Loschke, John Lewis, Jim Lesko, Alex Macalma.

Figure 137

Sikorsky JUH-60 Rotorcraft Aircrew Systems Concepts Airborne Laboratory (RASCAL).

A second Black Hawk, originally the Boeing Advanced Digital Optical Control System demonstrator, arrived in 1989 as the replacement vehicle for its predecessor, the CH-47B, to carry on the variable stability and control and guidance system research of the Center. This JUH-60A (NASA 750), now known as the Rotorcraft Aircrew Systems Concepts Airborne Laboratory (fig. 137) and dubbed RASCAL for short, is the most sophisticated in the long line of variable stability helicopters to be developed by NASA and the Army. The RASCAL was developed incrementally in a four-phase program led initially by Edwin Aiken, then by Bob Jacobsen, and Nick Rediess. While the advanced 32-bit Research Flight Control System (RFCS) was being produced under contract by Boeing for installation in phase 4, extensive vehicle and rotor-system instrumentation, a real-time stereo-video passive ranging system, and a sophisticated on-board image generation system (ref. 182) were developed in-house and used to conduct productive flight research. During this time Rediess and Ernie Moralez carried the responsibility for the RFCS-associated development, and Phil Smith developed the passive ranging system; Jay Fletcher and Eric Strassilla developed the hybrid laser/accelerometer-based blade-motion measurement system.

A principal focus of the work with the RASCAL helicopter is the development and evaluation of advanced flight control concepts to improve the agility of military rotorcraft, while also providing the pilot with carefree maneuvering within an automatically protected flight envelope. Other research applications include the development of active sensors (such as millimeter-wave radar), passive sensors (such as infrared), and symbologies for advanced displays. These are technologies needed to

Figure 138
RASCAL research team. Front row: Zsolt Halmos, Jim Ahlman, Sonya Mahal, Paul Aristo, Bob Brunelle, Brad Curelop, Chima Njaka. Second row: Jack Brilla, Shirley Worden-Burek, Paul Espinosa, Seth Kurasaki, Benny Cheung, Alla Silverman, Adel Delous, Sharon Cioffi, Larry Hintz, Trudy Schlaich, Ursula Hawkins, Janice Bachkosky, Amara DeKeczer, Ed Aiken, Tony Gudino. Third row: Zoltan Szoboszlay, Tom Kaisersatt, Bob Burney, Cas Lesiak, Ernie Moralez, Hossein Mansur, Jim Jeske, Gary Villere, Paul Everhart, John Foster, Bob Jacobsen, Luigi Cicolani, Vern Merrick, Rick Zelenka, Rich Coppenbarger, Bill Hindson. Fourth row: Court Bivens, Bill Decker, Mark Tischler, Stewart Anderson, Nick Rediess, Brian DeSilva, Mark Takahashi, Laura Iseler. Fifth row: Lee Mountz, Jack Trapp, Gary Leong, Roy Williams, K. C. Shih. Back row: Eric Strasilla, Amir Arani, Eric Weirshauser, Thad Frazier.

Figure 139
Bell JAH-1S FLITE Cobra.

Figure 140
Bell NAH-1S FLITE Cobra.

assist the military pilot in conducting nap-of-the-Earth flight at night and in adverse weather conditions; they can also improve the safety of civilian operations such as emergency medical services, fire fighting, or oil-rig support. The research team, which is carrying out a wide range of investigations, is shown in figure 138. As a part of the research into helicopter flight mechanics modeling, Jay Fletcher performed a thorough parameter identification for the aircraft in hover and at speeds of 40 and 80 knots in forward flight to define its characteristics for use in control law design (ref. 183). Bill Hindson conducted experiments to explore noise abatement approaches using differential global positioning system guidance (ref. 184). Richard Zelenka and Richard Coppenbarger carried out research on sensors and displays for low-altitude terrain flight (ref. 185). Hindson and George Tucker served as the principal pilots at the outset of the program. Currently, Rick Simmons and Army LTC Chris Sullivan perform that role. Based on Ames experience with nap-of-the-Earth guidance and control, Harry Swenson led an Ames team that carried out a flight program on the Army's STAR Black Hawk at Ft. Monmouth, New Jersey, to evaluate the use of a stored digital terrain base and flightpath-centered pursuit guidance for near terrain flight (ref. 186). Ames pilots on this program were Gordon Hardy and Munro Dearing.

From the start of Ames crew station and human factors flight research, experiments were carried out on the JAH-1S Cobra (fig. 139). This helicopter, called the Flying Laboratory for Integrated Test and Evaluation (FLITE), arrived at Ames in 1985. It was the first Cobra on which the prototype AH-64 visually coupled night vision system helmet mounted display (HMD) was installed. The aircraft took part in the Army's first use of visually coupled HMD systems and was later involved in a study of a modified communication system.[53] This system, designed by Zoltan Szoboszlay, allowed pilots to switch between three radios and internal communications with the co-pilot, without removing either hand from the flight control sticks (ref. 187). Loran Haworth was the principal investigator and project pilot. The aircraft was returned to the Army in 1985 for overhaul.

The NAH-1S (fig. 140), the successor to the original FLITE Cobra, has been used extensively in joint NASA/Army human factors research in the areas of night vision displays and voice communications since its arrival at Ames in 1987. It was originally modified for use as a surrogate trainer for pilots of the McDonnell Douglas AH-64 Apache helicopter through the installation of a pilot night vision system (PNVS). Haworth and Szoboszlay coordinated the test projects on the aircraft. In night-vision research, the aircraft was used for a study in which the pilot's performance was measured on a low-altitude course while the pilot's field-of-view was restricted to simulate that of night-vision devices. Human performance curves were generated as a function of field-of-view; the results were published in reference 188. In that same vein, the FLITE Cobra was also employed in two studies in which performance with 40-degree field-of-view night-vision goggles was compared with daytime performance with a 40-degree field-of-view restriction, and also with daytime performance without

[53] Loran Haworth 1998: personal communication.

this restriction (ref. 189). In another test, the effects of depth perception when using night-vision goggles and the PNVS FLIR (ref. 190) were examined. Visual estimation of altitude in near-terrain flight was performed and reported in reference 191. Concerning voice communications, pilots evaluated active noise-reduction technologies for better audio communications while piloting this helicopter (refs. 192 and 193). The aircraft was also used in tests of computer voice recognizers and synthesizers. In a different area of human factors research, the FLITE Cobra was used simultaneously with a fixed-base crew station simulator to conduct a simulation sickness study (ref. 194). Principal investigators for the various flight tests were Haworth and Szoboszlay, along with Army flight surgeon LTC John Crowley,

Figure 141
FLITE Cobra research team. Front row: Tom Reynolds, Nick Proett, Sean Hogan, Loran Haworth, John Browning. Second row: Mary Kaiser, John Spooner, Richard Lee, Munro Dearing, Sue Laurie, Paul Aristo, Alan Lee, Zsolt Halmos, Zoltan Szoboszlay, Dick Denman, Lee Mountz. Back row: David Foyle, Millard Edgerton, Ron Fong, Trudy Schlaich, Gary Leong, Linda Blyskal, Brian Hookland, Steve Timmons, Fran Kaster, Wendell Stephens, Alex Macalma, Dana Marcell.

David Foyle, Daniel Hart, Robert Hennessy, Thomas Sharkey, and Carol Simpson. Pilots who participated throughout were Haworth, Munro Dearing, Bill Hindson, George Tucker, and Army pilots LTC Thomas Reynolds, MAJ Ronald Seery, and LTC Rick Simmons. The FLITE team is shown in figure 141.

In addition to human factors studies, the FLITE Cobra was used to validate the military rotorcraft flying qualities specification maneuvers, visual cues, and test methods in a degraded visual environment. In support of the Army's RAH-66 Comanche Program, the aircraft was used as a radar target while hovering in front of a directed imaging radar, used for testing the F117 radar signature on the ground. This was the first use of the radar against a rotorcraft in hover flight. The tests proved that this type of technology radar is useful for measuring the radar signature of a rotorcraft through 360 degrees of rotation to test for radar-signature conformity. Also in support of the Comanche program, the aircraft flew prototype fiber-optic connectors to gather long-term fiber-optic attenuation data. In another test, a color video camera was installed and boresighted to the PNVS. Several hours of infrared and color video imagery were collected over various types of terrain for use in a part-task infrared trainer in a collaborative program with the Israeli Ministry of Defense. The aircraft is being modified to test an image fusion sensor for night vision, a programmable helmet-mounted display system, and an automated gearbox health-monitoring system.

A novel method of measuring rotorcraft impulsive noise in which a quiet aircraft is used as a microphone platform was developed by Fred Schmitz and his Army/NASA team in the mid-1970s. The aircraft was instrumented with a tail microphone and flown in formation with the test helicopter at selected airspeeds and rates of sink at which the helicopter was known to radiate large amounts of impulsive noise. The distance between the aircraft and the helicopter was held within 1-meter accuracy using a visual range finder. The concept was first proven by using a Grumman OV-1C Mohawk aircraft that was borrowed from the Army Engineering Flight Activity at Edwards Air Force Base.[54] The test team, led by Fred Schmitz and Donald Boxwell of the Army and supported by Army and NASA personnel, was the first to document and record true free-field measurements of helicopter impulsive noise. The results led to a new appreciation for and understanding of the major sources of these very complicated rotorcraft acoustic phenomena (ref. 195).

The success of this new measurement technique prompted a search for a better measurement platform, one that was quieter and better suited to the flight conditions in which rotorcraft impulsive noise was likely. The search led to a YO-3A that was originally built by Lockheed for the Army as a surveillance aircraft and operated by the Federal Bureau of Investigation. It was modified to accept on-board recording and monitoring equipment that included wing-tip microphones which, together with a tail microphone, helped assess the directivity of the radiated noise. The initial YO-3A team is shown in figure 142. Special emphasis was placed on quantitatively measuring

---

[54] Fred Schmitz 1998: personal communication.

Figure 143
Lockheed YO-3A.

Figure 142
Lockheed YO-3A acoustics research team. From left to right: Don Boxwell, Fred Schmitz, Bob Williams, Lee Jones, Bob George, Vance Duffy.

the impulsive noise during the landing approach of a series of Army helicopters (fig. 143); the results are reported in references 196 and 197. The aircraft was also used to evaluate the noise characteristics of the contending designs for the Army's Utility Tactical Transport Aircraft System and the Advanced Attack Helicopter, with results reported directly to the source selection board. The resulting external noise data proved invaluable and helped in selecting the eventual winners of both large Army contracts. This series of experiments also provided a unique database to the technical acoustic community that helped focus the research efforts of the next 20 years. These very successful in-flight acoustic test programs made it clear that this new research tool was an important asset to acoustic research.

In mid-1977, Ames acquired its own YO-3A and took up the role of providing in-flight acoustic measurements for future rotorcraft research programs. Upgraded equipment and instrumentation were added to the NASA YO-3A aircraft, but the basic measurement procedures remained similar to those of the earlier test programs. Use of the YO-3A by NASA for far-field measurement of rotorcraft noise is described in references 198 and 199. Several major research programs were undertaken by

NASA, with the Army as a major partner. Tests were carried out on the AH-1, XV-15, UH-60, S-76, BO-105, and MD-900 aircraft. Typical results from the flight research effort, including comparisons with data from tests in the 40- by 80-foot wind tunnel are presented in references 200 and 201.

The Ames flight operations staff of 1991 is shown in figure 144.

Figure 144
Flight operations personnel circa 1991. From left to right: Loran Haworth, Rick Simmons, Tom Reynolds, Goli Davidson, Jack McLaughlin, Trudy Schlaich, Jim Martin, Mike Stortz, Gordon Hardy, Patti Bergin, Nancy Lowe, Ron Seery, Larry Hintz, Mike Landis, Munro Dearing, George Tucker.

Epilogue

This history of flight research beginning at the NACA's Ames Aeronautical Laboratory in 1940 and continuing to this day at NASA's Ames Research Center has covered a wide range of technical areas that include icing research; transonic model testing; aerodynamics research; flying qualities, stability and control, and performance evaluations; variable stability aircraft; gunsight tracking and guidance and control displays; in-flight thrust reversing and steep approach research; boundary-layer control research; STOL and V/STOL aircraft research; and rotorcraft research. The flight research came about in many cases as the result of a progressive development of ideas through stages of analyses, wind tunnel, and ground-based simulator tests in Ames facilities. In fact, the collaborative efforts using this combination of facilities led to a more substantive result than would have been the case had the flight experiments been conducted in isolation. Several national awards were presented to Ames pilots and engineers in recognition of their achievements in pursuit of these research objectives. The significant contributions that came about as a result of these various programs are listed below.

• Prototypes of the Ames anti-icing system were evaluated in the B-17 and B-24 heavy bombers from the WWII era. A substantive program on the C-46 Commando icing research aircraft led to the definition of icing-system design criteria.

• Qualitative aerodynamic data in the transonic regime were obtained through tests of small wing-mounted and drop models.

• The effects of compressibility on aerodynamics, including pitch-up, shock-induced buffet, and aileron buzz were established.

• The YF-93A aircraft was the first to use flush NACA engine inlets and was tested extensively with both scoop and flush inlets.

• Drag characteristics of tailless delta winged aircraft were documented in flight and compared with wind tunnel data.

• Results from flight tests of the Ogee wing planform were provided to the Anglo-French team that was in the process of designing the Concorde supersonic transport, giving them assurance that the configuration was suitable for that aircraft.

• Data from wake-vortex penetration tests were generalized to form the basis for the FAA's separation criteria for aircraft landing behind large, heavy aircraft.

• Extensive results were obtained concerning the stability and control characteristics that influence the acceptable approach speed, providing an understanding of the selection of approach speed for high-performance aircraft.

• A standardized system for rating flying qualities was developed that accounted for the demands of the flight task and the behavior of the aircraft and pilot in accomplishing the task to the expected degree of precision. The Cooper-Harper rating scale has been one of the enduring contributions of flying-qualities research over the past 40 years.

• The world's first variable stability aircraft was developed and, along with several successors, contributed extensive data for use in defining flying qualities design criteria, as well as in assessing the flying qualities of numerous individual aircraft designs.

• A head-up guidance display was demonstrated at Ames in a Boeing 727-100 transport airplane and was subsequently adapted by the industry and certificated for civil transport operations.

• In-flight thrust reversing was evaluated with the F-94C Starfire fighter. The thrust-reversing concept was applied eventually to DC-8 transport aircraft to achieve the rapid descent capability required for commercial transport certification.

• A direct lift-control system was demonstrated to engineers and pilots of the airline industry. The design was incorporated as part of the Lockheed L-1011 commercial airliner's flight control system, and was particularly effective in achieving excellent automatic landing performance for that aircraft.

• Flight tests with the JF-104 Starfighter developed steep approach profiles now used by the space shuttle.

• Approach and landing tests provided basic design data for the space shuttle configuration control and the guidance and navigation systems, a demonstration of digital autopilot technology, and supported the decision to remove air-breathing engines from the final shuttle design.

• Extensive data on fully blown leading-edge and trailing-edge flaps were obtained in flight tests of the F-100A Super Sabre. The experience gained from this testing went into blown flap systems used on the F-104 Starfighter and the F-4 Phantom II.

• The Augmentor Wing Jet STOL Research Aircraft, the first jet STOL transport, was used in a joint NASA/Canadian Department of Industry, Trade and Commerce project to demonstrate the concept in the low-speed regime and in terminal-area operations. The aircraft was also used for evaluating flying qualities criteria, augmented controls, flight director concepts, and an automatic approach and landing system that allowed pilots to exploit the vehicle's unique STOL capabilities in terminal-area operations. The Augmentor Wing performed a flight demonstration of a nonlinear inverse control concept applied to a full flight envelope autopilot.

• The Quiet Short-Haul Research Aircraft (QSRA) design used upper-surface blowing to achieve short-field takeoff and landing performance. The aircraft documented stable flight at lift levels three times those generated on conventional aircraft and demonstrated operations aboard the aircraft carrier U.S.S. Kitty Hawk without need for launch catapults or landing arresting gear. The aircraft was also used to evaluate integrated flightpath/airspeed controls and displays for making precision instrument approaches and landings. The Air Force developed flying qualities specifications for the C-17 long-range, heavy transport based on the results of the earlier Ames Augmentor Wing and QSRA research.

• Extensive flight testing was carried out on the X-14 jet vertical takeoff and landing (VTOL) aircraft to investigate a range of flying qualities in hover and transition flight and to evaluate lateral thrust vectoring control. Research with this aircraft contributed to the military's V/STOL flying qualities specification and played an important role in the control system development of the Harrier prototype.

• The XV-15 tilt rotor was the first proof-of-concept vehicle built entirely to Ames' specifications. This aircraft became the subject of a series of technology development activities over the next decade. The work included flying qualities and stability and control evaluations, control law development, side-stick controller tests, performance evaluations in all flight modes, acoustics tests, flow surveys, and documentation of its loads, structural dynamics, and aeroelastic stability characteristics. The XV-15 provided the foundation necessary for the initiation of the V-22 Osprey program for the Marines and for current development of the Bell 609 civil tilt rotor.

• Research with the modified YAV-8B Harrier was used to develop flying qualities criteria and control system and display concepts for future STOVL fighter aircraft as part of the Joint Strike Fighter program. It also demonstrated display technology that is being implemented in the AV-8B Harrier II for shipboard operation. Its complex digital, fly-by-wire control system was designed, developed, and installed in the YAV-8B by an Ames team.

• The X-36 unmanned scale model tailless fighter demonstrated excellent stability and maneuverability up to stall angle of attack.

• The UH-1H helicopter was used to develop and evaluate control systems that would permit fully automatic flight for helicopters. The first demonstration of automatic control laws for a helicopter, developed using nonlinear inverse methods was conducted on the aircraft. Flying qualities experiments provided criteria used in the Army's helicopter flying qualities specification that was developed primarily at Ames, as well as control and display requirements for civil instrument flight operations.

• The CH-47B Chinook was used by the Army and NASA as a variable stability helicopter. The in-flight simulation capability permitted a wide variety of flight experiments that ranged from investigations in support of the Army's flying qualities specification to the evaluation of advanced multi-input, multi-output control laws.

• A UH-60 Black Hawk with sophisticated blade instrumentation was used to acquire extensive rotor-load data. Approximately 30 gigabytes of data were obtained and installed in an electronic database that was immediately accessible to the U.S. industry.

• A JUH-60A Black Hawk helicopter, known as RASCAL, was modified to incorporate extensive vehicle and rotor-system instrumentation, a digital research flight control system, a real-time stereo-video passive ranging system, and a sophisticated on-board image generation system. A principal focus of the RASCAL helicopter work is to develop advanced flight control designs to improve the agility of military rotorcraft, while also providing the pilot with carefree maneuvering within an automatically protected flight envelope.

• Two Cobra helicopters hosted Ames crew station and human factors flight research experiments. One of these aircraft has been used extensively in joint NASA/Army human factors research in the area of visual and auditory displays.

• A YO-3A aircraft, modified by the Army and NASA to allow accurate in-flight measurement of rotorcraft impulsive noise, contributed to the source selection of two Army helicopter programs. A second YO-3A was used extensively by NASA for rotorcraft acoustics research.

At this writing, most areas of flight research have once again been transferred to the NASA Dryden Flight Research Center. An exception to this directive has allowed rotorcraft flight research to continue at Ames with aircraft operated by the U.S. Army. Three aircraft will continue in the Ames inventory to carry on the flight research tradition at the Center. These are the two UH-60 Black Hawk helicopters (NASA 748 and 750) and the NAH-1S Cobra (NASA 736). Research in the future will be conducted as part of a joint NASA/Army rotorcraft technology program and will focus on advanced controls and cockpit interfaces with these aircraft. NASA 750 will continue to serve in the role as an in-flight simulator as well.

To conclude this history, we would like to recognize the three Ames aeronautical research pilots who lost their lives in carrying out their professional duties. Ryland Carter, Rudy Van Dyke, and Don Heinle were all highly accomplished pilots who were fearless in pursuit of their test objectives. They were respected and appreciated by their colleagues of the day and are remembered today as men who made significant contributions to aeronautical technology through their skill and daring.

Ames Research Center, Moffett Field, California, circa 1995

References

1. Hartman, Edwin P.: Adventures in Research: A History of Ames Research Center, 1940-1965. NASA SP-4302, 1970.

2. Rodert, Lewis A.; McAvoy, William H.; and Clousing, Lawrence A.: Preliminary Report on Flight Tests of an Airplane Having Exhaust Heated Wings. NACA Wartime Report A-53, 1941.

3. Look, Bonne C.: Flight Tests of the Thermal Ice-Protection Equipment on the B-17F Airplane. NACA Wartime Report A-23, 1944.

4. Selna, James; Neel, Carr B., Jr.; and Zeiller, E. Lewis: An Investigation of a Thermal Ice-Prevention System for a C-46 Cargo Airplane. IV. Results of Flight Tests in Dry-Air and Natural-Icing Conditions. NACA Wartime Report A-22, 1945.

5. Neel, Carr B., Jr.; and Bright, Loren G.: The Effect of Ice Formations on Propeller Performance. NACA TN-2212, 1950.

6. Jones, Alun R.: An Investigation of a Thermal Ice-Prevention System for a Twin-Engine Transport Airplane. NACA Report 862, 1946.

7. Rathert, George A., Jr.; Hanson, Carl M.; and Rolls, L. Stewart: Investigation of a Thin Straight Wing of Aspect Ratio 4 by the NACA Wing-Flow Method: Lift and Pitching Moment Characteristics of the Wing Alone. NACA RM-A8L20, 1949.

8. Rathert, George A., Jr.; and Cooper, George E.: Wing-Flow Tests of a Triangular Wing of Aspect Ratio Two. I. Effectiveness of Several Types of Trailing-Edge Flaps on Flat-Plate Models. NACA RM-A7G18, 1947.

9. Nissen, James M.; Gadeberg, Burnett L.; and Hamilton, William T.: Correlation of the Drag Characteristics of a Typical Pursuit Airplane Obtained from High-Speed Wind Tunnel and Flight Tests. NACA Report 916, 1948.

10. Clousing, Lawrence A.; and Turner, William N.: Flight Measurements of Horizontal Tail Loads on a Typical Propeller-Driven Pursuit Airplane during Stalled Pull-Outs at High Speed. NACA Wartime Report A-81, 1944.

11. Clousing, Lawrence A.; Turner, William N.; and Rolls, L. Stewart: Measurements in Flight of the Pressure Distribution on the Right Wing of a Pursuit-Type Airplane at Several Values of Mach Number. NACA TR-859, 1946.

12. Sadoff, Melvin; Turner, William N.; and Clousing, Lawrence A.: Measurements of the Pressure Distribution on the Horizontal-Tail Surface of a Typical Propeller Driven Pursuit Airplane in Flight: Effects of Compressibility in Steady Straight and Accelerated Flight. NACA TN-1144, 1947.

13. Turner, William N.; Steffen, Paul J.; and Clousing, Lawrence A.: Compressibility Effects on the Longitudinal Stability and Control of a Pursuit-Type Airplane As Measured in Flight. NACA Report 854, 1946.

14. White, Maurice D.; Lomax, Harvard; and Turner, Howard L.: Sideslip Angles and Vertical-Tail Loads in Rolling Pull-Out Maneuvers. NACA TN-1122, 1947.

15. Gasich, Welko E.; and Clousing, Lawrence A.: Flight Investigation of the Variation of Drag Coefficient with Mach Number for the Bell P-39N-1 Airplane. NACA Wartime Report A-61, 1945.

16. Cooper, George E.; and Rathert, George A., Jr.: Visual Observation of the Shock Wave in Flight. NACA RM-A8C25, 1948.

17. Cooper, George E.; and Bray, Richard S.: Schlieren Investigation of the Wing Shock-Wave Boundary-Layer Interaction in Flight. NACA RM-A51G09, 1951.

18. White, Maurice D.; Sadoff, Melvin; Clousing, Lawrence A.; and Cooper, George E.: Preliminary Results of a Flight Investigation to Determine the Effect of Negative Flap Deflection on High-Speed Longitudinal Control Characteristics. NACA RM-A7I26, 1947.

19. Gasich, Welko E.: Experimental Verification of Two Methods for Computing the Takeoff Ground Run of Propeller-Driven Aircraft. NACA TN 1258, 1947.

20. Clousing, Lawrence A.; Brown, Harvey H.; and Rathert, George A.: Flight-Test Measurements of Aileron Control Surface Behaviour at Supercritical Mach Numbers. NACA RM-A7A15, 1947.

21. Brown, Harvey H.; and Clousing, Lawrence A.: Wing Pressure-Distribution Measurements Up to 0.866 Mach Number in Flight on a Jet-Propelled Airplane. NACA TN-1181, 1947.

22. Spreiter, John R.; Galster, George M.; and Cooper, George E.: Flight Observations of Aileron Flutter at High Mach Number As Affected by Several Modifications. NACA RM-A7B03, 1947.

23. Clousing, Lawrence A.: Precautions for Flight Testing Near the Speed of Sound. NACA RM-A7G25, 1947.

24. Spreiter, John R.; and Steffan, Paul J.: Effect of Mach and Reynolds Numbers on Maximum Lift Coefficient. NACA TN-1044, 1946.

25. Anderson, Seth B.; and Bray, Richard S.: A Flight Evaluation of the Longitudinal Stability Characteristics Associated with the Pitch-Up of a Swept-Wing Airplane in Maneuvering Flight at Transonic Speeds. NACA-TR-1237, 1955.

26. McFadden, Norman M.; and Heinle, Donovan R.: Flight Investigation of the Effects of Horizontal-Tail Height, Moment of Inertia, and Control Effectiveness on the Pitch-Up Characteristics of a 35 deg Swept-Wing Fighter Airplane at High Subsonic Speeds. NACA RM-A54F21, 1955.

27. Anderson, Seth B.; Matteson, Frederick H.; and Van Dyke, Rudolph D., Jr.: A Flight Investigation of the Effect of Leading-Edge Camber on the Aerodynamic Characteristics of a Swept-Wing Airplane. NACA RM-A52L16a, 1953.

28. Matteson, Frederick H.; and Van Dyke, Rudolph D., Jr.: Flight Investigation of the Effects of a Partial-Span Leading-Edge Chord Extension on the Aerodynamic Characteristics of a 35 deg Swept-Wing Fighter Airplane. NACA RM-A54B26, 1954.

29. Sadoff, Melvin; Stewart, John D.; and Cooper, George E.: Analytical Study of Comparative Pitch-Up Behavior of Several Airplanes and Correlation with Pilot Opinion. NACA RM-A57D04, 1957.

30. Rathert, George A., Jr.; Winograd, Lee; Cooper, George E.; and Rolls, L. Stewart: Preliminary Flight Investigation of the Wing-Dropping Tendency and Lateral Control Characteristics of a 35 deg Swept-Wing Airplane at Transonic Mach Numbers. NACA RM-A50H03, 1950.

31. McFadden, Norman M.; Rathert, George A., Jr.; and Bray, Richard S.: The Effectiveness of Wing Vortex Generators in Improving the Maneuvering Characteristics of a Swept-Wing Airplane at Transonic Speeds. NACA TN-3523, 1955.

32. Bray, Richard S.: The Effects of Fences on the High-Speed Longitudinal Stability of a Swept-Wing Airplane. NACA RM-A53F23, 1953.

33. Rolls, L. Stewart: A Flight Comparison of a Submerged Inlet and a Scoop Inlet at Transonic Speeds. NACA RM-A53A06, 1953.

34. Rolls, L. Stewart; Havill, C. Dewey; and Holden, George R.: Techniques for Determining Thrust In-Flight for Airplanes Equipped with Afterburners. NACA RM-A52K12, 1953.

35. Rolls, L. Stewart; and Wingrove, Rodney C.: An Investigation of the Drag Characteristics of a Tailless Delta-Wing Airplane in Flight, Including Comparison with Wind-Tunnel Data. NACA Memo 10-8-58A, Nov. 1958.

36. Rolls, L. Stewart; Koenig, David G.; and Drinkwater, Fred J., III: Flight Investigation of the Aerodynamic Properties of an Ogee Wing. NASA TN D-3071, 1965.

37. Jacobsen, Robert A.; and Short, Barbara J.: A Flight Investigation of the Wake Turbulence Alleviation Resulting from a Flap Configuration Change on a B-747 Aircraft. NASA TM-73,263, 1977.

38. Tinling, Bruce E.: Estimation of Vortex-Induced Roll Excursions Based on Flight and Simulation Results. In: FAA-RD-77-68 Proceedings of the Aircraft Wake Vortices Conference, Mar. 1977, pp. 11–22.

39. Short, Barbara J.; and Jacobsen, Robert A.: Evaluation of a Wake Vortex Upset Model Based on Simultaneous Measurements of Wake Velocities and Probe-Aircraft Accelerations. NASA TM-78561, 1979.

40. Turner, Howard L.; and Cooper, George E.: Partial Measurements in Flight of the Flying Qualities of a Grumman XF7F-1 Airplane with a Modified Vertical Tail. NACA RM-A7D15, 1947.

41. Spahr, J. Richard; and Christopherson, Don R.: Measurements in Flight of the Stability, Lateral Control, and Stalling Characteristics of an Airplane Equipped with Full Span Flaps and Spoiler-Type Ailerons. NACA Wartime Report A-28, Dec. 1943.

42. Hanson, Carl M.; and Anderson, Seth B.: Flight Tests of a Double-Hinged Horizontal Tail Surface with Reference to Longitudinal-Stability and -Control Characteristics. NACA TN-1224, 1947.

43. Cole, Henry A., Jr.; and Holleman, Euclid C.: Measured and Predicted Dynamic Response Characteristics of a Flexible Airplane to Elevator Control over a Frequency Range Including Three Structural Modes. NACA TN-4147, 1958.

44. Cole, Henry A., Jr.; Brown, Stuart C.; and Holleman, Euclid C.: Experimental and Predicted Longitudinal and Lateral-Directional Response Characteristics of a Large Flexible 358 Swept-Wing Airplane at an Altitude of 35,000 Feet. NACA Report 1330, 1957.

45. Goett, Harry J.; Jackson, Roy P.; and Belsley, Steven E.: Wind-Tunnel Procedure for Determination of Critical Stability and Control Characteristics of Airplanes. NACA Report 781, 1944.

46. Delany, Noel K.; and Kauffman, William M.: Correlation of Wind-Tunnel Predictions with Flight Tests of a Twin-Engine Patrol Airplane. I. Longitudinal-Stability and -Control Characteristics. NACA Wartime Report A-86, 1945.

47. Delany, Noel K.; and Kauffman, William M.: Correlation of Wind-Tunnel Predictions with Flight Tests of a Twin-Engine Patrol Airplane. II. Lateral- and Directional-Stability and -Control Characteristics. NACA Wartime Report A-71, 1945.

48. Anderson, Seth B.; Christofferson, Frank E.; and Clousing, Lawrence A.: Flight Measurement of the Flying Qualities of a Lockheed P-80A Airplane: Longitudinal Stability and Control Characteristics. NACA RM-A7G01, 1947.

49. Anderson, Seth B.; and Cooper, George E.: Flight Measurements of the Flying Qualities of a Lockheed P-80A Airplane: Lateral- and Directional-Stability and -Control Characteristics. NACA RM-A7J24, 1947.

50. Heinle, Donovan R.; and McNeill, Walter E.: Correlation of Predicted and Experimental Lateral Oscillation Characteristics for Several Airplanes. NACA RM-A52J06, 1952.

51. Anderson, Seth B.: Correlation of Flight and Wind-Tunnel Measurements of Roll-Off in Low-Speed Stalls on a 35 deg Swept-Wing Aircraft. NACA RM-A53G22, 1953.

52. Sadoff, Melvin; Matteson, Frederick H.; and Van Dyke, Rudolph D., Jr.: The Effect of Blunt-Trailing-Edge Modifications on the High-Speed Stability and Control Characteristics of a Swept-Wing Fighter Airplane. NACA RM-A54C31, 1954.

53. McNeill, Walter E.; and Cooper, George E.: A Comparison of the Measured and Predicted Lateral Oscillatory Characteristics of a 35 deg Swept-Wing Fighter Airplane. NACA TN-3521, 1955.

54. Anderson, Seth B.; and Bray, Richard S.: A Flight Evaluation of the Longitudinal Stability Characteristics Associated with the Pitch-Up of a Swept-Wing Airplane in Maneuvering Flight at Transonic Speeds. NACA Report 1237, 1955.

55. Sadoff, Melvin: The Effects of Longitudinal Control-System Dynamics on Pilot Opinion and Response Characteristics as Determined from Flight Tests and from Ground Simulator Studies. NASA Memo 10-1-58A, 1958.

56. White, Maurice D.; and Innis, Robert C.: A Flight Investigation of Low Speed Handling Qualities of a Tailless Delta-Wing Fighter Airplane. NASA Memo 4-15-59A, May 1959.

57. McNeill, Walter E.; and Innis, Robert C.: A Simulator and Flight Study of Yaw Coupling in Turning Maneuvers of Large Transport Aircraft. NASA TN D-3910, 1967.

58. White, Maurice D.; Schlaff, Bernard A.; and Drinkwater, Fred J., III: A Comparison of Flight-Measured Carrier-Approach Speeds with Values Predicted by Several Different Criteria for 41 Fighter-Type Airplane Configurations. NACA RM-A57L11, 1958.

59. Drinkwater, Fred J., III; and Cooper, George E.: A Flight Evaluation of the Factors Which Influence the Selection of Landing Approach Speeds. NASA Memo 10-6-58A, Dec. 1958.

60. Innis, Robert C.: Factors Limiting the Landing Approach Speed of Airplanes from the Viewpoint of a Pilot. AGARD Report 358, 1961.

61. Creer, Brent Y.; Stewart, John D.; Merrick, Robert B.; and Drinkwater, Fred J., III: Pilot Opinion Study of Lateral Control Requirements for Fighter-Type Aircraft. NASA Memo 1-29-59A, Mar. 1959.

62. Cooper, George E.: Understanding and Interpreting Pilot Opinion. Aeronautical Engineering Review, vol. 16, no. 3, Mar. 1957, pp. 47–51.

63. Cooper, George E.; and Harper, Robert P., Jr.: The Use of Pilot Rating in the Evaluation of Aircraft Handling Qualities. NASA TN D-5153, 1969.

64. Kauffman, William M.; Liddell, Charles J., Jr.; Smith, G. Allan; and Van Dyke, Rudolph D., Jr.: An Apparatus for Varying Effective Dihedral in Flight with Application to a Study of Tolerable Dihedral on a Conventional Fighter Airplane. NACA Report 948, 1949.

65. McNeill, Walter E.; and Creer, Brent Y.: A Summary of Results Obtained during Flight Simulation of Several Aircraft Prototypes with Variable Stability Airplanes. NACA RM-A56C08, 1956.

66. Liddell, Charles J., Jr.; Van Dyke, Rudolph D., Jr.; and Heinle, Donovan R.: A Flight Determination of the Tolerable Range of Effective Dihedral on a Conventional Fighter Airplane. NASA TN-1936, 1949.

67. Liddell, Charles J., Jr.; Creer, Brent Y.; and Van Dyke, Rudolph D., Jr.: A Flight Study of Requirements for Satisfactory Lateral Oscillatory Characteristics of Fighter Aircraft. NACA RM-A51E16, 1951.

68. McNeill, Walter E.; Drinkwater, Fred J., III; and Van Dyke, Rudolph D., Jr.: A Flight Study of the Effects on Tracking Performance of Changes in the Lateral-Oscillatory Characteristics of a Fighter Airplane. NACA RM-A53H10, 1953.

69. McNeill, Walter E.; and Vomaske, Richard F.: A Flight Investigation to Determine the Lateral Oscillatory Damping Acceptable for an Airplane in the Landing Approach. NASA Memo 12-10-58A, Feb. 1959.

70. Vomaske, Richard F.; Sadoff, Melvin; and Drinkwater, Fred J., III: The Effect of Lateral-Directional Control Coupling on Pilot Control of an Airplane as Determined in Flight and a Fixed-Base Flight Simulator. NASA TN D-1141, 1961.

71. McFadden, Norman M.; Pauli, Frank A.; and Heinle, Donovan R.: A Flight Study of Longitudinal Control-System Dynamic Characteristics by the Use of a Variable Control-System Airplane. NACA RM-A57L10, 1958.

72. McFadden, Norman M.; Vomaske, Richard F.; and Heinle, Donovan R.: Flight Investigation Using Variable Stability Airplanes of Minimum Stability Requirements for High-Speed, High-Altitude Vehicles. NASA TN D-779, 1961.

73. Kauffman, William M.; and Drinkwater, Fred J., III: Variable Stability Airplanes in Lateral Stability Research. Aeronautical Engineering Review, vol. 14, no. 8, Aug. 1955, pp. 29–35.

74. Foster, John V.: Servomechanisms as Used on Variable-Stability and Variable-Control-System Research Aircraft. Proceedings of the National Electronics Conference, Oct. 1957, pp. 167–177.

75. McNeill, Walter E.; Gerdes, Ronald M.; Innis, Robert C.; and Ratcliff, Jack D.: A Flight Study of the Use of Direct-Lift-Control Flaps to Improve Station Keeping during In-Flight Refueling. NASA TM X-2936, 1973.

76. Rathert, George A., Jr.; Gadeberg, Burnett L.; and Ziff, Howard L.: An Analysis of the Tracking Performances of Two Straight-Wing and Two Swept-Wing Fighter Aircraft with Fixed Sights in Standardized Test Maneuver. NACA RM-A53H12, 1953.

77. Gadeberg, Burnett L.; and Rathert, George A., Jr.: Tracking Performance of a Swept-Wing Fighter With a Disturbed Reticle Lead-Computing Sight. NACA RM-A54K16, 1959.

78. Rathert, George A., Jr.; Abramovitz, M.; and Gadeberg, B. L.: The Effects of Powered Controls and Fire Control Systems on Tracking Accuracy. NACA RM-A55D12a, 1955.

79. Doolin, Brian; Smith, G. Allan; and Drinkwater, Fred J., III: An Air-Borne Target Simulator for Use in Optical-Sight Tracking Studies. NACA RM-A55F20, 1955.

80. Douvillier, Joseph G., Jr.; Foster, John V.; and Drinkwater, Fred J., III: An Airborne Simulator Investigation of the Accuracy of an Optical Track Command Missile Guidance System. NACA RM-A56G24, 1956.

81. Foster, John V.; Fulcher, Elmer C.; and Heinle, Donovan R.: An Airborne Target Simulator for Use with Scope-Presentation Type Fire-Control Systems. NACA RM-A57C19, 1957.

82. Turner, Howard L.; White, John S.; and Van Dyke, Rudolph D., Jr.: Flight Testing by Radio Remote Control-Flight Evaluation of a Beep-Control System. NACA TN-3496, 1955.

83. Turner, Howard L.; Triplett, William C.; and White, John S.: A Flight and Analog Computer Study of Some Stabilization and Command Networks for an Automatically Controlled Interceptor during the Final Attack Phase. NACA RM-A54J14, 1955.

84. McNeill, Walter E.; McLean, John D.; Hegarty, Daniel M.; and Heinle, Donovan R.: Design and Flight Tests of an Adaptive Control System Employing Normal-Acceleration Command. NASA TN D-858, 1961.

85. Kibort, Bernard R.; and Drinkwater, Fred J., III: A Flight Study of Manual Blind Landing Performance Using Closed Circuit Television Displays. NASA TN D-2252, 1964.

86. Bray, Richard S.: A Head-Up Display Format for Application to Transport Aircraft Approach and Landing. NASA TM-81199, 1980.

87. McGee, Leonard A.; Smith, Gerald L.; Hegarty, Daniel M.; Carson, Thomas M.; Merrick, Robert B.; Schmidt, Stanley F.; and Conrad, Bjorn: Flight Results from a Study of Aided Inertial Navigation Applied to Landing Operations. NASA TN D-7302, 1973.

88. Lee, Homer Q.; Newman, Frank; and Hardy, Gordon H.: 4D Area Navigation System Description and Flight Test Results. NASA TN D-7874, 1975.

89. Erzberger, Heinz: Automation of On-Board Flightpath Management. NASA TM-84212, 1981.

90. Denery, Dallas G.; Callas, George P.; Hardy, Gordon H.; and Nedell, William: Design, Development and Flight Test of a Demonstration Advanced Avionics System. AGARD Conference Proceedings No. 34, 1983.

91. Paielli, Russell A.; Bach, Ralph E.; McNally, B. David; Simmons, Rickey C.; Warner, David N.; Forsyth, Theodore J.; Kanning, Gerd; Ahtye, C. T.; Kaufmann, D. N.; and Walton, J. C.: Carrier Phase Differential GPS Integrated with an Inertial Navigation System: Flight Test Evaluation with Auto-Coupled Precision Landing Guidance. Proceedings of The Institute of Navigation's 1995 National Technical Meeting, Jan. 1995, pp. 711–724.

92. Triplett, William C.; Brown, Stuart C.; and Smith, G. Allan: The Dynamic-Response Characteristics of a 35 deg Swept-Wing Airplane as Determined from Flight Measurements. NACA Report 1250, 1955.

93. Denery, Dallas G.: An Identification Algorithm That Is Insensitive to Initial Parameter Estimates. AIAA Journal, vol. 9, no. 3, Mar. 1971, pp. 371–377.

94. Bach, Ralph E., Jr.; and Wingrove, Rodney C.: Applications of State Estimation in Aircraft Flight Data Analysis. AIAA-83-2087, Aug. 1983.

95. Tischler, Mark B.: System Identification Methods for Aircraft Flight Control Development and Validation. NASA TM-110369, Oct. 1995.

96. Anderson, Seth B.; Cooper, George E.; and Faye, Alan, E., Jr.: Flight Measurements of the Effect of a Controllable Thrust Reverser on the Flight Characteristics of a Single-Engine Jet Airplane. NASA Memo 4-26-59A, 1959.

97. Bray, Richard S.; Snyder, C. Thomas; and Drinkwater, Fred J., III: A Preliminary Flight and Simulator Study of the Use of Wing Spoilers for Direct Lift Control in the Approach and Landing of a Jet Transport Airplane. Ames Research Center Working Paper 224.

98. Bray, Richard S.; Drinkwater, Fred J., III; and White, Maurice D.: A Flight Study of a Power-Off Landing Technique Applicable to Re-Entry Vehicles. NASA TN D-323, July 1960.

99. Edwards, Fred G.; Foster, John D.; Hegarty, Daniel M.; Smith, Donald W.; Drinkwater III, Fred J.; and Wingrove, Rodney C.: Flight Performance of a Navigation, Guidance and Control System Concept for Automatic Approach and Landing of the Space Shuttle Orbiter. NASA TN D-7899, 1975.

100. Edwards, Fred G.; Foster, John D.; Hegarty, Daniel M.; and Drinkwater III, Fred J.: Delayed Flap Approach Procedures for Noise Abatement and Fuel Conservation. Proceedings of the NASA Aircraft Safety and Operating Problems Conference, Oct. 1976, pp. 77–90.

101. Denery, Dallas G.; Bourquin, Kent R.; White, Kenneth C.; and Drinkwater, Fred J. III: Flight Evaluation of Three-Dimensional Area Navigation for Jet Transport Noise Abatement. Journal of Aircraft, vol. 10, no. 4, Apr. 1973, pp. 226–231.

102. Holzhauser, Curt A.; and Bray, Richard S.: Wind Tunnel and Flight Investigations of the Use of Leading-Edge Area Suction for the Purpose of Increasing the Maximum Lift Coefficient of a 35 deg Swept-Wing Airplane. NACA Report 1276, 1956.

103. Cook, Woodrow; Anderson, Seth B.; and Cooper, George E.: Area-Suction Boundary-Layer Control as Applied to the Trailing-Edge Flaps of a 35 deg Swept-Wing Airplane. NACA Report 1370, 1958.

104. Kelly, Mark W.; Anderson, Seth B.; and Innis, Robert C.: Blowing-Type Boundary-Layer Control as Applied to the Trailing-Edge Flaps of a 35 deg Swept-Wing Airplane. NACA Report 1369, 1958.

105. Quigley, Hervey C.; Anderson, Seth B.; and Innis, Robert C.: Flight Investigation of the Low-Speed Characteristics of a 45 deg Swept-Wing Fighter-Type Airplane with Blowing Boundary-Layer Control Applied to the Leading- and Trailing-Edge Flaps. NASA TN D-321, 1960.

106. Rolls, L. Stewart; Cook, Anthony M.; and Innis, Robert C.: Flight-Determined Aerodynamic Properties of a Jet-Augmented, Auxiliary-Flap, Direct Lift-Control System Including Correlation with Wind-Tunnel Results. NASA TN D-5128, 1969.

107. Quigley, Hervey C.; and Innis, Robert C.: Handling Qualities and Operational Problems of a Large Four-Propeller STOL Transport Airplane. NASA TN D-1647, 1963.

108. Quigley, Hervey C.; Innis, Robert C.; Vomaske, Richard F.; and Ratcliff, Jack W.: A Flight and Simulator Study of Directional Augmentation Criteria for a Four-Propellered STOL Airplane. NASA TN D-3909, 1967.

109. Quigley, Hervey C.; Innis, Robert C.; and Holzhauser, Curt A.: A Flight Investigation of the Performance, Handling Qualities, and Operational Characteristics of a Deflected Slipstream STOL Transport Airplane Having Four Interconnected Propellers. NASA TN D-2231, 1964.

110. Innis, Robert C.; Holzhauser, Curt A.; and Gallant, Richard P.: Flight Tests under IFR with a STOL Transport Aircraft. NASA TN D-4939, 1968.

111. Holzhauser, Curt A.; Innis, Robert C.; and Vomaske, Richard F.: A Flight and Simulator Study of the Handling Qualities of a Deflected Slipstream STOL Seaplane Having Four Propellers and Boundary-Layer Control. NASA TN D-2966, 1965.

112. Innis, Robert C.; Holzhauser, Curt A.; and Quigley, Hervey C.: Airworthiness Considerations for STOL Aircraft. NASA TN D-5594, 1970.

113. Few, David D.: A Perspective on 15 Years of Proof of Concept Aircraft Development and Flight Research at Ames-Moffett by the Rotorcraft and Powered-Lift Flight Projects Division, 1970-1985. NASA Reference Publication 1187, 1987.

114. Weiberg, James A.; Giulianetti, Demo; Gambucci, Bruno; and Innis, Robert C.: Takeoff and Landing Performance and Noise Characteristics of a Deflected Slipstream STOL Airplane with Interconnected Propellers and Rotating Cylinder Flaps. NASA TM X-62,320, 1973.

115. Quigley, Hervey C.; Innis, Robert C.; and Grossmith, S.: A Flight Investigation of the STOL Characteristics of an Augmented Jet Flap STOL Research Aircraft. NASA TM X-62,334, 1974.

116. Vomaske, Richard F.; Innis, Robert C.; Swan, Brian E.; and Grossmith, Seth W.: A Flight Investigation of the Stability, Control, and Handling Qualities of an Augmented Jet Flap STOL Airplane. NASA TP-1254, 1978.

117. Stephenson, Jack D.: The Application of Parameter Estimation to Flight Measurements to Obtain Lateral-Directional Stability Derivatives of an Augmented Jet-Flap STOL Airplane. NASA TP-2033, 1983.

118. Franklin, James A.; Innis, Robert C.; Hardy, Gordon H.; and Stephenson, Jack D.: Design Criteria for Flightpath and Airspeed Control for the Approach and Landing of STOL Aircraft. NASA TP-1911, 1982.

119. Hindson, William S.; Hardy, Gordon H.; and Innis, Robert C.: Flight-Test Evaluation of STOL Control and Flight Director Concepts in a Powered-Lift Aircraft Flying Curved Decelerating Approaches. NASA TP-1641, 1981.

120. Watson, DeLamar M.; Hardy, Gordon H.; and Warner, David N., Jr.: Flight Test of the Glide-Slope Track and Flare-Control Laws for an Automatic Landing System for a Powered-Lift STOL Airplane. NASA TP-2128, 1983.

121. Meyer, George; and Cicolani, Luigi S.: Application of Nonlinear Systems Inverses to Automatic Flight Control Design: Systems Concepts and Flight Evaluations. In: AGARDograph 10, July 1981.

122. Cochrane, John A.; Riddle, Dennis W.; Stevens, Victor C.; and Shovlin, Michael D.: Selected Results from the Quiet Short-Haul Research Aircraft Flight Research Program. Journal of Aircraft, vol. 19, no. 12, Dec. 1982, pp. 1076–1082.

123. Stevens, Victor C.; Riddle, Dennis W.; Martin, James L.; and Innis, Robert C.: Powered-Lift STOL Aircraft Shipboard Operations: A Comparison of Simulation, Land-Based and Sea-Trials Results for the QSRA. AIAA Paper 81-2480, 1981.

124. Riddle, Dennis W.; Innis, Robert C.; Martin, James L.; and Cochrane, John A.: Powered-Lift Takeoff Performance Characteristics Determined from Flight Test of the Quiet Short-Haul Research Aircraft (QSRA). AIAA Paper 81-2409, Nov. 1981.

125. Stephenson, Jack D.; and Hardy, Gordon H.: Longitudinal Stability and Control Characteristics of the Quiet Short-Haul Research Aircraft (QSRA). NASA TP-2965, 1989.

126. Stephenson, Jack D.; Jeske, James A.; and Hardy, Gordon H.: Lateral-Directional Stability and Control Characteristics of the Quiet Short-Haul Research Aircraft (QSRA). NASA TM-102250, 1990.

127. Franklin, James A.; Hynes, Charles S.; Hardy, Gordon H.; Martin, James L.; and Innis, Robert C.: Flight Evaluation of Augmented Controls for Approach and Landing of Powered-Lift Aircraft. Journal of Guidance, Control, and Dynamics, vol. 9, no. 5, Sept.–Oct. 1986, pp. 555–565.

128. Hynes, Charles S.; Franklin, James A.; Hardy, Gordon H.; Martin, James L.; and Innis, Robert C.: Flight Evaluation of Pursuit Displays for Precision Approach of Powered-Lift Aircraft. Journal of Guidance, Control, and Dynamics, vol. 12, no. 4, July–Aug. 1989, pp. 521–529.

129. Hynes, Charles S.; Hardy, Gordon H.; and Kaisersatt, Thomas J.: Flight Evaluation of an Integrated Control and Display System for Precision Manual Landing Flare of Powered-Lift STOL Aircraft. SAE Proceedings of the 1987 International Powered Lift Conference, Paper P-203, Dec. 1987, pp. 165–180.

130. Watson, DeLamar M.; Hardy, Gordon H.; Innis, Robert C.; and Martin, James L.: Flight Evaluation of a Precision Landing Task for a Powered-Lift STOL Aircraft. AIAA-86-2130, Aug. 1986.

131. Cochrane, John A.; Riddle, Dennis W.; and Stevens, Victor C.: Quiet Short-Haul Research Aircraft: The First Three Years of Flight Research. AIAA-81-2625, Dec. 1981.

132. Riddle, Dennis W.; Stevens, Victor C.; and Eppel, Joseph C.: Quiet Short-Haul Research Aircraft: A Summary of Flight Research Since 1981. SAE Technical Paper 872315, Dec. 1987.

133. Anderson, Seth B.: Handling Qualities Criteria for V/STOL Aircraft. AGARDograph 46, pt. 2, June 1960, pp. 511–531.

134. Drinkwater, Fred J., III: Operational Technique for Transition of Several Types of V/STOL Aircraft. NASA TN D-774, 1961.

135. Turner, Howard L.; and Drinkwater, Fred J., III: Some Flight Characteristics of a Deflected Slipstream V/STOL Aircraft. NASA TN D-1891, 1963.

136. Pauli, Frank A.; Hegarty, Daniel M.; and Walsh, Thomas M.: A System for Varying the Stability and Control of a Deflected-Jet Fixed-Wing VTOL Aircraft. NASA TN D-2700, 1965.

137. Rolls, L. Stewart; and Drinkwater, Fred J., III: A Flight Determination of the Attitude Control Power and Damping Requirements for a Visual Hovering Task in the Variable Stability and Control X-14A Research Vehicle. NASA TN D-1328, 1962.

138. Rolls, L. Stewart; Drinkwater, Fred J., III; and Innis, Robert C.: Effects of Lateral Control Characteristics on Hovering a Jet Lift VTOL Aircraft. NASA TN D-2701, 1965.

139. Gerdes, Ronald M.; and Weick, Richard F.: A Preliminary Piloted Simulator and Flight Study of Height Control Requirements for VTOL Aircraft. NASA TN D-1201, 1962.

140. Feistel, Terrell W.; Gerdes, Ronald M.; and Fry, Emmett B.: An Investigation of a Direct Side-Force Maneuvering System on a Deflected Jet VTOL Aircraft. NASA TN D-5175, 1969.

141. Corliss, Lloyd D.; Greif, Richard K.; and Gerdes, Ronald M.: Comparison of Ground-Based and In-Flight Simulation of VTOL Hover Control Concepts. Journal of Guidance and Control, vol. 1, no. 3, May–June 1978, pp. 217–221.

142. Holzhauser, Curt A.; Morello, Samuel A.; Innis, Robert C.; and Patton, James M., Jr.: A Flight Evaluation of a VTOL Jet Transport Under Visual and Simulated Instrument Conditions. NASA TN D-6754, 1972.

143. Quigley, Hervey C.; and Koenig, David G.: A Flight Study of the Dynamic Stability of a Tilting-Rotor Convertiplane. NASA TN D-778, 1961.

144. Dugan, Daniel C.; Erhart, Ronald G.; and Schroers, Laurel G.: The XV-15 Tilt Rotor Research Aircraft. NASA TM-81244, 1980.

145. Maisel, Martin D.; and Harris, D. J.: Hover Tests of the XV-15 Tilt Rotor Research Aircraft. AIAA-81-2501, 1981.

146. Churchill, Gary B.; and Gerdes, Ronald M.: Advanced AFCS Developments on the XV-15 Tilt Rotor Research Aircraft. Proceedings of the American Helicopter Society's 40th Annual Forum, May 1984.

147. Schroers, Laurel G.: Dynamic Structural Aeroelastic Stability Testing of the XV-15 Tilt Rotor Research Aircraft. NASA TM-84293, 1982.

148. Marr, Roger L.; Blackman, Sheppard; Weiberg, James A.; and Schroers, Laurel G.: Wind Tunnel and Flight Test of the XV-15 Tilt Rotor Research Aircraft. Proceedings of the American Helicopter Society's 35th Annual Forum, May 1979.

149. Cook, Woodrow L.; and Hickey, David H.: Correlation of Low Speed Wind Tunnel and Flight Test Data for V/STOL Aircraft. NASA TM X-62,423, 1975.

150. Gerdes, Ronald M.; and Hynes, Charles S.: Factors Affecting Handling Qualities of a Lift-Fan Aircraft during Steep Terminal Area Approaches. NASA TM X-62,424, 1975.

151. Gerdes, Ronald M.: Lift-Fan Aircraft: Lessons Learned The Pilot's Perspective. NASA CR-177620, 1993.

152. Foster, John D; Moralez, Ernesto III; Franklin, James A.; and Schroeder, Jeffery A.: Integrated Control and Display Research for Transition and Vertical Flight on the NASA V/STOL Research Aircraft (VSRA). NASA TM-100029, 1987.

153. Franklin, James A.; Stortz, Michael W.; Borchers, Paul F.; and Moralez, Ernesto III: Flight Evaluation of Advanced Controls and Displays for Transition and Landing on the NASA V/STOL Systems Research Aircraft. NASA TP-3607, 1996.

154. Dorr, Daniel W.; Moralez, Ernesto III; and Merrick, Vernon K.: Simulation and Flight Test Evaluation of Head-Up-Display Guidance for Harrier Approach Transitions. AIAA Paper 92-4233, 1992.

155. Borchers, Paul F.; Moralez, Ernest III; Merrick, Vernon K.; and Stortz, Michael W.: YAV-8B Reaction Control System Bleed and Control Power Usage in Hover and Transition. NASA TM-104021, 1994.

156. Birckelbaw, Larry D.; and Nelson, E. L.: Infrared Flow Visualization of V/STOL Aircraft. AIAA Paper 92-4253, 1992.

157. Gea, L.; Chyu, W. J.; Stortz, Michael W.; Roberts, Andrew C.; and Chow, C.: Flight Test and Numerical Simulation of Transonic Flow Around YAV-8B Harrier II Wing. AIAA Paper 91-1628, 1991.

158. Foster, John D.; McGee, Leonard A.; and Dugan, Daniel C.: Helical Automatic Approaches of Helicopters with Microwave Landing Systems. NASA TP-2109, 1982.

159. Swenson, Harry N.; Hamlin, J. R.; and Wilson, Grady W.: NASA-FAA Helicopter Microwave Landing System Curved Path Flight Test. Proceedings of the American Helicopter Society's 40th Annual Forum, May 1984, pp. 447–459.

160. Corliss, Lloyd D.; and Carico, Dean G.: A Helicopter Flight Investigation of Roll-Control Sensitivity, Damping, and Cross-Coupling in a Low Altitude Maneuvering Task. NASA TM-84376, 1983.

161. Lebacqz, J. Victor: NASA/FAA Experiments Concerning Helicopter IFR Airworthiness Criteria. NASA TM-84388, 1983.

162. Cross, Jeffrey L.; and Watts, Michael E.: Tip Aerodynamics and Acoustics Test: A Report and Data Survey. NASA RP-1179, 1988.

163. Watts, Michael E.; Cross, Jeffrey L.; and Noonan, K. W.: Two-Dimensional Aerodynamic Characteristics of the OLS/TAAT Airfoil. NASA TM-89435, 1988.

164. Flemming, Robert J.; and Erickson, Ruben E.: An Evaluation of Vertical Drag and Ground Effect Using the RSRA Rotor Balance System. Proceedings of the American Helicopter Society's 38th Annual Forum, June 1982, pp. 43–54.

165. Erickson, Ruben E.; Kufeld, Robert M.; Cross, Jeffrey L.; Hodge, R. W.; Ericson, William F.; and Carter, Robert D. G.: NASA Rotor System Research Aircraft Flight-Test Data Report: Helicopter and Compound Configuration. NASA TM-85843, 1984.

166. Acree, Cecil W., Jr.: Preliminary Report on In-Flight Measurement of Rotor Hub Drag and Lift Using the RSRA. NASA TM-86764, 1985.

167. Erickson, Ruben E.; Cross, Jeffrey L.; Kufeld, Robert M.; Acree, Cecil W., Jr.; Nguyen, D.; and Hodge, R. W.: NASA Rotor Systems Research Aircraft: Fixed-Wing Configuration Flight-Test Results. NASA TM-86789, 1986.

168. Watson, Douglas C.; and Hindson, William S.: In-Flight Simulation Investigation of Rotorcraft Pitch-Roll Cross Coupling. NASA TM-101059, 1988.

169. Hindson, William S.; Tucker, George E.; Lebacqz, J. Victor; and Hilbert, Kathryn B.: Flight Evaluation of Height Response Characteristics for the Hover Bob-Up Task and Comparison with Proposed Criteria. Proceedings of the American Helicopter Society's 42nd Annual Forum, June 1986, pp. 685–709.

170. Chen, Robert T. N.; and Tischler, Mark B.: The Role of Modeling and Flight Testing in Rotorcraft Parameter Identification. Vertica, vol. 11, no. 4, 1987, pp. 619–646.

171. Hilbert, Kathryn B.; Lebacqz, J. Victor; and Hindson, William S.: Flight Investigation of a Multivariable Model-Following Control System for Rotorcraft. AIAA Paper 86-9779, 1986.

172. Chen, Robert T. N.; and Hindson, William S.: Analytical and Flight Investigation of the Influence of Rotor and Other High-Order Dynamics on Helicopter Flight-Control System Bandwidth. NASA TM-86696, 1985.

173. Holdridge, Richard D.; Hindson, William S.; and Bryson, Arthur E.: LQG-Design and Flight Test of a Velocity-Command System for a Helicopter. Proceedings of the AIAA Guidance and Control Conference, Aug. 1985.

174. Eshow, Michelle M.: Flight Investigation of Variations in Rotorcraft Control and Display Dynamics for Hover. Journal of Guidance, Control, and Dynamics, vol. 15, no. 2, Mar.–Apr. 1992, pp. 482–490.

175. Schroeder, Jeffery A.; and Merrick, Vernon K.: Flight Evaluations of Several Hover Control and Display Combinations for Precise Blind Vertical Landings. AIAA-90-3479, 1990.

176. Tischler, Mark B.; Fletcher, Jay W.; Morris, Patrick M.; and Tucker, George E.: Flying Quality Analysis and Flight Evaluation of Highly Augmented Combat Rotorcraft. Journal of Guidance, Control, and Dynamics, vol. 14, no. 5, Sept.–Oct. 1991, pp. 954–963.

177. Cross, Jeffrey L.; Brilla, John M.; Kufeld, Robert M.; and Balough, Dwight L.: The Modern Rotor Aerodynamic Limits Survey: A Report and Data Survey. NASA TM-4446, 1993.

178. Kufeld, Robert M.; Balough, Dwight L.; Cross, Jeffrey L.; Studebaker, Karen E.; Jennison, Christopher D.; and Bousman, William G.: Flight Testing the UH–60A Airloads Aircraft. Proceedings of the American Helicopter Society's 50th Annual Forum, May 1994, pp. 557–578.

179. Mueller, Arnold W.; Conner, David A.; Rutledge, Charles K.; and Wilson, Mark R.: Full Scale Flight Acoustic Results for the UH-60A Airloads Aircraft. Proceedings of the American Helicopter Society's Vertical Lift Aircraft Design Conference, Paper No. 5, Jan. 1995.

180. Bousman, William G.: A Qualitative Examination of Dynamic Stall from Flight Test Data. Proceedings of the American Helicopter Society's 53rd Annual Forum. April–May 1997, pp. 368–387.

181. Cicolani, Luigi; McCoy, Allen; Tischler, Mark; Tucker, George; Gatenio, P.; and Marmar, D.: Flight-Time Identification of a UH-60A Helicopter and Sling Load. Paper No. 10, AGARD/RTA Symposium on System Identification for Integrated Aircraft Development and Flight Testing, May 1998.

182. Jacobsen, Robert A.; Rediess, Nicholas A.; Hindson, William S.; Aiken, Edwin W.; and Bivens, C. C.: Current and Planned Capabilities of the NASA/Army Rotorcraft Aircrew Systems Concepts Airborne Laboratory (RASCAL). Proceedings of the American Helicopter Society's 51st Annual Forum, May 1995, pp. 758–771.

183. Fletcher, Jay W.: Identification of UH-60 Stability Derivative Models in Hover from Flight Test Data. Journal of American Helicopter Society, vol. 40, no. 1, 1995, pp. 32–46.

184. Hindson, William S.; and Chen, Robert T. N.: Flight Tests of Noise Abatement Approaches for Rotorcraft Using Differential GPS Guidance. AHS, Annual Forum, 51st, Fort Worth, TX, May 9–11, 1995; Proceedings. Vol. 1; Alexandria, VA, American Helicopter Society, 1995, pp. 681–694.

185. Zelenka, Richard E.; Smith, Phillip N.; Coppenbarger, Richard A.; Njaka, Chima E.; and Sridhar, Banavar: Results from the NASA Automated Nap-of-the-Earth Program. AHS, Annual Forum, 52nd, Washington, DC, June 4–6, 1996; Proceedings. Vol. 1; Alexandria, VA, American Helicopter Society, 1996, pp. 197–209.

186. Swenson, Harry N.; Zelenka, Richard E.; Dearing, Munro G.; Hardy, Gordon H.; Clark, Raymond; Davis, Tom; Amatrudo, Gary; and Zirkler, Andre: Design and Flight Evaluation of an Integrated Navigation and Near-Terrain Helicopter Guidance System for Night-Time and Adverse Weather Operations. NASA TM-108837, 1994.

187. Haworth, Loran A.; Szoboszlay, Zoltan P.; Shively, Robert J.; and Bick, Frank J.: AH-1S Communications Switch Integration Program. NASA TM-101053, 1989.

188. Haworth, Loran A.; Szoboszlay, Zoltan P.; Kasper, Eugene F.; DeMaio, Joe; and Halmos, Zsolt L.: In-flight Simulation of Visionic Field-of-View Restrictions on Rotorcraft Pilot's Workload, Performance and Visual Cueing. Proceedings of the American Helicopter Society's 52nd Annual Forum, May 1996, pp. 1155–1185.

189. Szoboszlay, Zoltan P.; Edwards, Kenneth; Haworth, Loran A.; Pratty, Adam; White, John; and Halmos, Zsolt L.: Predicting Usable Field-of-View Limits for Future Rotorcraft Helmet-Mounted Display. Proceedings of Innovation in Rotorcraft Technology, The Royal Aeronautical Society, Paper No. 7, 1997.

190. Foyle, David C.; and Kaiser, Mary K.: Pilot Distance Estimation with Unaided Vision, Night-Vision Goggles and Infrared Imagery. SID International Symposium Digest of Technical Papers, vol. XXII, May 1991, pp. 314–317.

191. Crowley, John; Haworth, Loran A.; Szoboszlay, Zoltan P.; and Lee, Alan G.: Helicopter Pilot Estimation of Self-Altitude in a Degraded Visual Environment. Proceedings of the Workshop on the Validation of Measurements, Models, and Theories, TTCP/HUM/97/006. The Technical Cooperation Program, Subcommittee on Non-Atomic Military Research and Development, Monterey, California, 10 June 1995.

192. Chan, Jeffrey W.; and Simpson, Carol A.: Comparison of Speech Intelligibility in Cockpit Noise Using SPH-4 Flight Helmet with and without Active Noise Reduction. NASA-CR-177564, July 1990.

193. Simpson, Carol; and King, Robert: Active Noise Reduction Flight Tests in Military Helicopters. In: AGARD Conference Proceedings 596, Paper No. 22, Oct. 1996.

194. Sharkey, Thomas J.; Matsumoto, Joy A.; Hennessy, Robert T.; and Voorhees, J. W.: Simulator-Induced Alteration of Head Movements. Proceedings of the AIAA/AHS Flight Simulation Technologies Conference, Aug. 1992, pp. 29–36.

195. Schmitz, Fredric H.; and Boxwell, Donald A.: In-Flight Far-Field Measurement of Helicopter Impulsive Noise. Journal of the American Helicopter Society, vol. 21, no. 4, Oct. 1976, pp. 2–16.

196. Boxwell, Donald A.; and Schmitz, Fredric H.: Full-Scale Measurements of Blade-Vortex Interaction Noise. Journal of the American Helicopter Society, vol. 27, no. 4, Oct. 1982, pp. 11–27.

197. Boxwell, Donald A.; and Schmitz, Fredric H.: In-Flight Comparison of the 540 and K747 Main Rotors for the AH-1G Helicopter: Production Validation Test. USAAEFA Report No. 77-38, Oct. 1979.

198. Cross, Jeffrey L.; and Watts, Michael E.: In-Flight Acoustic Testing Techniques Using the YO-3A Acoustic Research Aircraft. NASA TM-85895, 1984.

199. McCluer, Megan S.; and Dearing, Munro G., III: Measuring Blade-Vortex Interaction Noise Using The YO-3A Acoustics Research Aircraft. Proceedings of the 22nd European Rotorcraft Forum, Paper No. 82, Sept. 1996.

200. Yamauchi, Gloria K.; Signor, David B.; Watts, Michael E.; Hernandez, F. J.; and LeMasurier, P.: Flight Measurements of Blade-Vortex Interaction Noise Including Comparisons with Full-Scale Wind Tunnel Data. Proceedings of the American Helicopter Society's 49th Annual Forum, Vol. 1, May 1993, pp. 25–53.

201. Kitaplioglu, C.; McCluer, Megan S.; and Acree, Cecil W., Jr.: Comparison of XV-15 Full-Scale Wind Tunnel and In-Flight Blade-Vortex Interaction Noise. Proceedings of the American Helicopter Society's 53rd Annual Forum, Vol. 3, May 1997, pp. 153–163.

Index